CCCC STUDIES IN WRITING & RHETORIC

Edited by Victor Villanueva, Washington State University

The aim of the CCCC Studies in Writing & Rhetoric Series is to influence how we think about language in action and especially how writing gets taught at the college level. The methods of studies vary from the critical to historical to linguistic to ethnographic, and their authors draw on work in various fields that inform composition—including rhetoric, communication, education, discourse analysis, psychology, cultural studies, and literature. Their focuses are similarly diverse—ranging from individual writers and teachers, to work on classrooms and communities and curricula, to analyses of the social, political, and material contexts of writing and its teaching.

SWR was one of the first scholarly book series to focus on the teaching of writing. It was established in 1980 by the Conference on College Composition and Communication (CCCC) in order to promote research in the emerging field of writing studies. As our field has grown, the research sponsored by SWR has continued to articulate the commitment of CCCC to supporting the work of writing teachers as reflective practitioners and intellectuals.

We are eager to identify influential work in writing and rhetoric as it emerges. We thus ask authors to send us project proposals that clearly situate their work in the field and show how they aim to redirect our ongoing conversations about writing and its teaching. Proposals should include an overview of the project, a brief annotated table of contents, and a sample chapter. They should not exceed 10,000 words.

To submit a proposal, please register as an author at www.editorialmanager.com/nctebp. Once registered, follow the steps to submit a proposal (be sure to choose SWR Book Proposal from the drop-down list of article submission types).

On Multimodality

New Media in Composition Studies

Jonathan Alexander
University of California, Irvine

Jacqueline Rhodes
California State University, San Bernardino

Conference on College Composition and Communication

National Council of Teachers of English

Staff Editor: Bonny Graham
Series Editor: Victor Villanueva
Interior Design: Mary Rohrer
Cover Design: Mary Rohrer and Lynn Weckhorst

NCTE Stock Number: 34122; eStock Number: 34139
ISBN 978-0-8141-3412-2; eISBN 978-0-8141-3413-9

It is the policy of NCTE in its journals and other publications to provide a forum for the open discussion of ideas concerning the content and the teaching of English and the language arts. Publicity accorded to any particular point of view does not imply endorsement by the Executive Committee, the Board of Directors, or the membership at large, except in announcements of policy, where such endorsement is clearly specified.

Every effort has been made to provide current URLs and email addresses, but because of the rapidly changing nature of the Web, some sites and addresses may no longer be accessible.

Publication partially funded by a subvention grant from the Conference on College Composition and Communication of the National Council of Teachers of English.

Library of Congress Cataloging-in-Publication Data

Alexander, Jonathan, 1967-

On multimodality : new media in composition studies / Jonathan Alexander, University of California, Irvine ; Jacqueline Rhodes, California State University, San Bernardino.

pages cm. — (CCCC Studies in Writing & Rhetoric.)

Includes bibliographical references and index.

ISBN 978-0-8141-3412-2 (pbk.)

1. English language—Rhetoric—Computer-assisted instruction. 2. Mass media—Authorship—Study and teaching. 3. Online data processing—Authorship—Study and teaching. 4. English language—Rhetoric—Study and teaching—Data processing. 5. Report writing—Study and teaching—Data processing. 6. Report writing—Computer-assisted instruction. 7. Computers and literacy. 8. Multimedia systems. 9. Modality (Linguistics) I. Rhodes, Jacqueline, 1965– II. Title.

PE1404.A544 2014

808'.0420285—dc23

2013049766

CONTENTS

Permission Acknowledgments vii

Acknowledgments ix

Introduction 1

1. Refiguring Our Relationship to New Media 28
2. Direct to Video: Rewriting the Literacy Narrative 70
3. Prosumerism, Photo Manipulation, and Queer Spectacle 105
4. Collaboration, Interactivity, and the *Dérive* in Computer Gaming 127
5. Theorizing the Multimodal Subject 171

Notes 203

Works Cited 211

Index 223

Authors 231

PERMISSION ACKNOWLEDGMENTS

Images and text from *Viewmaster* installation reprinted from "Installation, Instantiation, and Performance" by Jacqueline Rhodes and Jonathan Alexander in *College Composition and Communication Online* 1.1 (January 2012). Used with permission.

Material from Kat Eason's Rhetoric in Practice project reprinted with permission.

Material from Elizabeth Losh's "Digital Rhetoric" course reprinted with permission. For more information, see https://eee.uci.edu/08f/25823.

Parts of Chapter 3 originally appeared in "Queerness: An Impossible Subject for Composition" by Jonathan Alexander and Jacqueline Rhodes in *JAC: A Journal of Rhetoric, Culture, and Politics* 31.1–2 (2011), as well as in "Queerness, Multimodality, and the Possibilities of Re/Orientation" in *Composing (Media) = Composing (Embodiment): Bodies, Technologies, Writing, the Teaching of Writing* edited by Kristin L. Arola and Anne Frances Wysocki, published by Utah State University Press (2012). Used with permission.

Initial arguments in Chapter 4 originally published in "Gaming, Student Literacies, and the Composition Classroom: Some Possibilities for Transformation" by Jonathan Alexander in *College Composition and Communication* 61.1 (September 2009). Used with permission.

"Technologies of the Self in the Aftermath: Affect, Subjectivity, and Composition," by Jonathan Alexander and Jacqueline Rhodes, in *Rhetoric Review* (2010) reprinted by permission of Taylor & Francis Ltd., http://www.tandf.co.uk/journals.

ACKNOWLEDGMENTS

We'd like to first thank Mack McCoy and Aurora Wolfgang. They know why.

Any book results from many conversations with many people. We are fortunate to have had some amazing conversations about our ideas with people both at home and afar. Locally, we have learned much from colleagues such as Mary Boland, Carol Burke, Kimberly Costino, Kat Eason, Loren Eason, Ellen Gil-Gómez, Daniel M. Gross, Lynda Haas, Monika Hogan, Susan Jarratt, David Lacy, Elizabeth Losh, Bonnie Nardi, Renée Pigeon, and Elaina Taylor. At more of a distance, we've greatly appreciated conversations with Linda Adler-Kassner, Kristin Arola, Brian Bailie, Kris Blair, Lillian Bridwell-Bowles, Theresa Enos, Jenn Fishman, Bre Garrett, Robin Gosser, Bump Halbritter, Wayne Hall, Gail Hawisher, Will Hochman, Mary Hocks, Deborah Holdstein, Karen Lunsford, Laura Micciche, Jason Palmeri, Eileen Schell, Cindy Selfe, Michael Spooner, Carl Whithaus, Lynn Worsham, and Anne Wysocki.

We have benefited tremendously from the peer review and feedback from those who have responded to earlier versions and manifestations of the arguments we explore in this book. We would like to point readers to a fuller discussion of the *Viewmaster* installation, mentioned in the introduction, in volume 1.1 of *College Composition and Communication Online* (2012). We acknowledge gratefully that parts of Chapter 3 originally appeared in "Queerness: An Impossible Subject for Composition," which was published in *JAC* 31.1–2 (2011) and "Queerness, Multimodality, and the Possibilities of Re/Orientation," a chapter in *Composing (Media) = Composing (Embodiment): Bodies, Technologies, Writing, the Teaching of Writing* edited by Kristin L. Arola and Anne Frances Wysocki for Utah State University Press, 2012. Initial arguments in Chapter 4 first came out in "Gaming, Student Literacies, and the Composition Classroom: Some Possibilities for Transformation," published in *College Composition and Communication* 61.1 (September 2009). And finally, discussions in Chapter 5 originated in "Technologies of the Self in the Aftermath: Affect, Subjectivity, and Composition," published in *Rhetoric Review* 29.2 (2010). We owe continuing debts to the reviewers and editors who first worked with us on developing and refining our arguments, as well as the reviewers for the complete manuscript of *On Multimodality*.

Finally, we owe a large debt to Bonny Graham, our most excellent editor at NCTE, and to Victor Villanueva, who directed our attention patiently and expertly to gaps in argument, flaws in reasoning, and missteps in rhetorical awareness. Any remaining gaps, flaws, and missteps are purely our own.

Introduction

IN 2009, CYNTHIA L. SELFE PUBLISHED AN essay in *College Composition and Communication* (*CCC*), "The Movement of Air, the Breath of Meaning," that both galvanized the move in our field to embrace multimodal and multimedia compositional practices and articulated the potential consequences for our disciplinarity in a way that provoked immediate attention and debate. Focusing specifically on the prevalence of sound as a modality of communicating experience, exploring insights, and representing identity, Selfe criticizes the lack of attention in composition courses to sound as a communicative domain. Noting the increase in attention to the visual field as well as multimodal composition in general, Selfe articulates what's "at stake," as she puts it, in "broadening" the domain of communicative practices that composition courses engage:

> As faculty, when we limit our understanding of composing and our teaching of composition to a single modality, when we focus on print alone as the communicative venue for our assignments and for students' responses to those assignments, we ensure that instruction is less accessible to a wide range of learners, and we constrain students' ability to succeed by offering them an unnecessarily narrow choice of semiotic and rhetorical resources. By broadening the choice of composing modalities, I argue, we expand the field of play for students with different learning styles and differing ways of reflecting on the world; we provide the opportunity for them to study, think critically about, and work with new communicative modes. (644)

Selfe's impassioned argument, grounded in her understanding of what the literacy education of contemporary US college students most needs, created immediate buzz when it appeared. The Writing Program Administrator's (WPA) listserv picked up the issue for discussion, and Kathleen Blake Yancey, editor of *CCC*, hosted virtual forums to debate Selfe's claims for expanding the modalities considered by composition.

Doug Hesse's response to Selfe's essay, published in a subsequent issue of *CCC*, argues quite clearly that Selfe may advocate for too much:

> My purpose is to temper Selfe's thoughtful argument, because the practices it advocates entail more than some supplemental tweak of current courses. At stake are fundamental boundaries of our curricular landscape and our sense of its stakeholders, interests, and purposes. Like many readers of her article, I'm inclined to make a place for the aurality in my own teaching even as I ponder whether I'm overstepping or sidestepping professional roles that best serve student and social interests. (605)

The "battle lines" could not be more clearly drawn; and indeed, Hesse's metaphoric evocation of boundaries and landscapes gestures toward agon—the agon of a discipline that seems increasingly divided. Selfe's advocacy of bringing yet another medium of communication into the composition classroom, in this case sound, signaled to some that our discipline has perhaps become a bit too open, and that we should—if in fact we *are* a discipline—be more discreet in proclaiming our proper objects of study and expertise. Viewing the situation in another way, however, advocates for Selfe's position agreed with her that rhetorical practices are in fact the proper domain of composition studies. As Selfe puts it in her article,

> Composition classrooms can provide a context not only for *talking about* different literacies, but also for *practicing* different literacies, learning to create texts that combine a range of modalities as communicative resources: exploring their af-

> fordances, the special capabilities they offer to authors; identifying what audiences expect of texts that deploy different modalities and how they respond to such texts. ("Movement" 643; author's emphasis)

What is at stake in such arguments is a fundamental consideration (and reconsideration) of what we as compositionists do.

This book argues that composition studies has found itself at a crossroads. In our steady incorporation of new media and multimedia forms of composing into our curricula and pedagogies, we have begun to meet the challenges of expanded notions of authoring, composing, and literacy. At the same time, the field's incorporation of new media into the composition classroom has often proceeded, we believe, by refiguring diverse multimedia as forms of "authoring" and the creation of different "texts." Students, for instance, "write" multimodal "texts" that "argue" points. While such work often enlivens and energizes both faculty and students, particularly as it builds on student interest in new media technologies, this work has tended to treat new media as though they were actually *new*. In the process, the multiple rhetorical capabilities of media—sound, visuals, video, and other multimodal media—have been elided, and the rich histories of those capabilities with them. In our push to assert our own disciplinarity, we have perhaps privileged text-based forms of writing to the extent that we rarely address the specific invention, delivery, and rhetorical possibilities of other types of composition in our classes.

In her introduction to Bedford/St. Martin's "critical sourcebook" on multimodal composition, which collects important articles and chapters on the subject, Claire Lutkewitte offers a tentative definition of *multimodal composition* as "communication using multiple modes that work purposely to create meaning" (2). We appreciate this definition, which leaves much room for considering media-specific rhetorical affordances, diverse modes and platforms of content creation and delivery, and a multiplicity of communicative strategies. Lutkewitte notes that "multimodal composition is not simply an extension of traditional composition, and we can't simply overlay traditional frameworks onto composing with multiple modes" (3). We couldn't agree more. But Lutkewitte, in an effort perhaps to bridge composition with (and into) multimodality, argues, "at the same time . . . the entire field of composition exists because of evolving theory and practice—multimodal composition included" (3). Our concern focuses on that evolution, on the fits and starts, the push and pull, the steps forward and backward as composition grapples with what it means to engage in, support, and study multimodal composing. On the one hand, we want to agree with Lutkewitte that composition will evolve, in theory and practice, as it engages multimodality; on the other hand, we fear that composition often just "includes" the multimodal, co-opting it as an "extension of traditional composition," as opposed to exploring how multimodality challenges our rhetorical predispositions in privileging print textualities.

Many in the subfield of computers and composition studies will readily understand this argument. For instance, Anne Frances Wysocki and Johndan Johnson-Eilola ask the question pointedly in their groundbreaking essay, "Blinded by the Letter: Why Are We Using Literacy as a Metaphor for Everything Else?" For Wysocki and Johnson-Eilola, the key question for consideration when thinking about composing with multimedia is, "What other possibilities might we use for expressing our relationships with and within technologies?" (349). In this book, we hope to further that questioning. As composition as a discipline embraces technology and actively invites students in first-year and advanced composition courses to compose with new and multimedia, we need to ask about other possibilities for expression, for representation, for communicating meaning, for making knowledge. We need to ask about possibilities that may exceed those of the letter, the text-based, the author, the *composed*. For to be sure, as multimedia have their own rich and varied histories, they have their own distinct modes, logics, methods, processes, and capabilities. A failure to "pay attention" to technologies (to borrow from Selfe's 1999 *Technology and Literacy in the 21st Century: The Importance of Paying Attention*) may delimit our ability to work productively with and for our students' multiple literacy needs. We seek to extend Selfe's challenge by asserting that a failure to pay attention to the specific rhetorical and production capabilities of new and multimedia may hamper our ability to understand the challenges that multimedia bring to understanding "literacy" and communicative possibilities in the twenty-first century. We may fail to meet our students' most pressing needs as communicators.

Of course, addressing the multiple dimensions of multiple media that far exceed the provenance of the letter will almost certainly mean that our much-prized disciplinarity will give way to increased inter- and multidisciplinarity. The communications challenges and possibilities of our time require a significantly deeper understanding of "composing" than our discipline currently offers. Obviously, trying to figure the "field" as one thing, a monolithic entity, would be ludicrous. Diversity—of method, of inquiry, of practice, of pedagogy—serves as one of the great intellectual pleasures (and challenges) of composition studies. We are many things, fulfilling many different functions, missions, and purposes. Jason Palmeri, in his 2012 book *Remixing Composition: A History of Multimodal Writing Pedagogy*, notes how writing instructors have for decades asked students to engage in multimodal writing, long before the widespread use of computers in the classroom. Teachers and students have worked together on visually rich magazine articles, sound essays, performance-driven compositions, and a variety of other "analog" pieces that are nonetheless richly multimodal. Palmeri advocates for a "pluralist vision" of composition studies that acknowledges how our field, at least since the 1960s, has always been engaged with multimodality, so the leap to multimedia should not be a large one, conceptually. Perhaps, however, the leap is larger than we think. Evidence remains that composition may not quite yet be meeting the challenge of incorporating multimodal and multimedia into its understanding of itself.

While we acknowledge and appreciate the extensiveness of Palmeri's account of multimodal practices in the teaching of writing, at least in US universities of the past four to five decades, we question the extent to which such practices have had an impact on our conception of what we do. Palmeri himself admits that "first-year composition programs will need to continue to place a special emphasis on alphabetic writing," even if "transforming curricula" to include at least "one formal [multimodal] assignment sequence" seems a "reasonable" request (152). The emphasis on the alphabetic finds resonance in the recent (2009) first edition of *The Norton Book of Composition Studies* (S. Miller), which contains very few articles about the use of multimodal practices or multimedia in the teaching of writing and instead focuses on the history of the development of the field and practices in the teaching of print-based writing. In the 1,800-page anthology, only five essays substantively address multimodal or multimedia writing: Selfe's "Technology and Literacy: The Perils of Not Paying Attention"; Diana George's "From Analysis to Design: Visual Communication in the Teaching of Writing"; Carolyn Miller and Dawn Shepherd's "Blogging as Social Action: A Genre Analysis of the Weblog"; Susan Romano, Barbara Field, and Elizabeth De Huergo's "Web Literacies of the Already Accessed and Technically Inclined: Schooling in Monterey, Mexico"; and Cynthia Selfe, Gail Hawisher, Dipo Lashore, and Pengfei Song's "Literacies and the Complexities of the Global Digital Divide." Granted, these are important contributions, but many of them are increasingly dated, focusing on simple calls to include computers in the teaching of writing and not to overlook the visual field as relevant to producing text-driven essays. Two of the pieces pull double duty with international student issues, figuring the Internet as a space for building job-related competencies. Indeed, when we as compositionists have adopted new and multimedia, we have often done so with print- and text-driven rhetorics in mind. We frequently use new media to foster the development of print-based literacies, or we fail to recognize the particular rhetorical affordances of multimedia and the communicative possibilities of thinking about "writing" in terms of design, visuality, and aurality. As authors, we do not exempt ourselves from this critique. In an early essay, for instance, Jonathan advocated for the use of "e-zines" in the writing classroom to help students develop a sense of audience; he theorized at the time that "posting student writing to the Web may be essentially equivalent to publishing class booklets" (Alexander, "Digital Spins" 388). In retrospect, such a statement seems ridiculously naive, particularly given the multimedia composing possibilities to which many of our students have access.

We agree with Palmeri that alphabetic writing and textual production should still be taught. And certainly, many compositionists have been thinking critically and pedagogically about different "literacies," particularly the visual. Palmeri's and Selfe's work alone attests to a recognition from some in our field that our students' compositional landscape has changed. However, we would like to further the discussion by arguing that our students' increasing personal engagement with multimedia technologies, as well as the steady adoption of multimedia in the classroom, challenges the primacy of alphabetic writing in how we communicate. David Sheridan, Jim Ridolfo, and Anthony Michel argue in "The Available Means of Persuasion: Mapping a Theory and Pedagogy of Multimodal Public Rhetoric" that, in the current "transformation of rhetorical education," the "academy's privileging of the written word; the cultural logics that circumscribe the use of certain modes, media, and technologies; and the division of rhetorical labor—[all] would be exposed for scrutiny. These dynamics would be reconceived in terms of the needs and goals of public rhetorics" (813). We believe that composition studies—at least as represented in the canon-forming text of our field's *Norton* anthology—has yet to acknowledge and grapple with what such a challenge means for how we teach writing, authoring, and composing.

How have we come to this view? Curiously, or perhaps not so curiously, through some of our own experimentations with using multimedia and technology, both in the composition classroom and as part of our own composing and scholarly processes. In the remainder of this introduction, we consider some of the experiences that have brought us to write this book, detail some of our critiques of how composition incorporates new media into its courses and curricula, and outline some possibilities for reimagining composition as a multidiscipline that takes seriously the challenges of new media's many rhetorical affordances.

AN INVITATION TO A VIEW: THE DILEMMAS OF TECHNO-COMP

At the 2009 computers and writing conference (C&W), hosted at the University of California, Davis, we mounted a multimedia installation entitled *Viewmaster.* As longtime members in the subfield of computers and composition and as frequent presenters at past C&W conferences, we knew how open and generous our colleagues could be in helping us think through the use of new media platforms as compositional spaces, and we felt that the conference was perhaps the best venue through which to experiment with multimedia forms of composing and critique. During the conference's exhibition session, interested individuals entered the *Viewmaster* room, a generously provided separate and enclosed space, in which viewers saw three large screens upon which were projected large and rather ominous pairs of eyes (our own, in fact) interspersed with a series of questions and short texts from Michel Foucault, Luis Buñuel, and the 2007 "Conference on College Composition and Communication (CCCC) Statement on the Multiple Uses of Writing." Many who saw *Viewmaster* stayed for several cycles of the installation's movement through the looming pairs of eyes and the alternating texts. Some fled the room nearly immediately, finding the eyes too grotesque.

We couldn't help but be pleased with both responses.

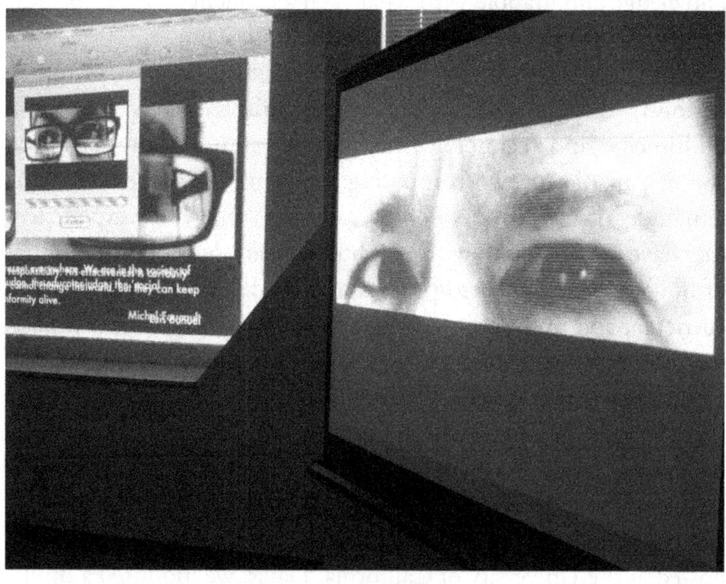

But beyond our pleasure at inducing affective responses (of nearly any kind), we were pleased that viewers engaged us in conversation about the piece, particularly as discussions revolved around the two topics we were most interested in exploring. First, visitors wanted to discuss the conglomeration of texts, questions, and visuals as a form of intellectual inquiry. The texts were simple. On the central screen, the text of the CCCC statement cycled through regularly:

> The CCCC hereby calls together—and calls to action—all those who share its vision of a future in which an expansive writing curriculum, backed by ample resources, attends unyieldingly to the difficult work of helping students use good words, images, and other appropriate means, well composed, to build a better world. ("Statement")

Following this statement appeared a series of alternating questions:

> Whose vision is this?
> What do they see?
> What do they want to see?
> What are good words?
> What are good images?
> What—or who—is well composed?
> Who sees a better world?

A pair of eyes appeared in between cycles of the statement and questions, while the two adjacent screens featured the ominous pairs of eyes with, first, a quotation from Foucault—"The judges of normality are present everywhere. We are in the society of the teacher-judge, the doctor-judge, the educator-judge, the 'social worker'-judge" (*Discipline* 304)—and second, from Buñuel—"In any society, the artist has a responsibility. His effectiveness is certainly limited and a painter or writer cannot change the world. But they can keep an essential margin of non-conformity alive" (Fuentes 93).

With this mixing and matching of simple visuals and pointed texts, we intended *Viewmaster* to provoke consideration of the often unremarked, frequently unacknowledged pressures that sur-

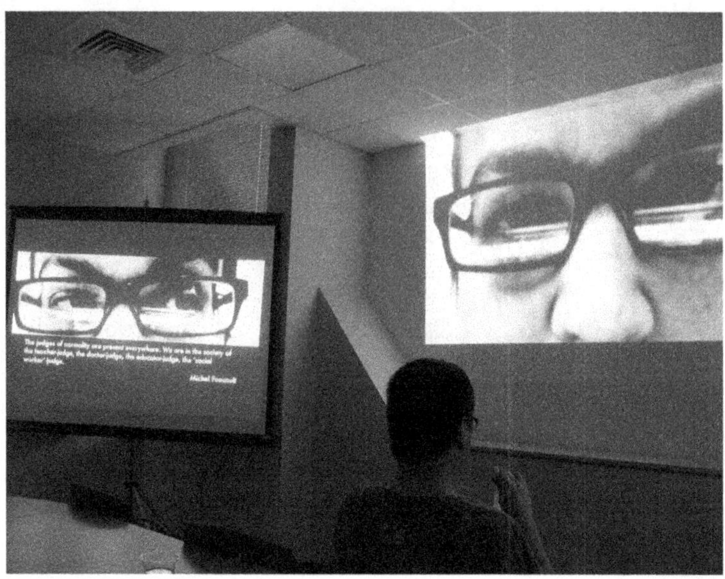

round the act of composition. Beginning with "CCCC Statement on the Multiple Uses of Writing," we wanted to tease out questions about our discipline's collective vision for writing, about the use of "appropriate means," and about what unnamed ideologies present themselves in our use of words and images to "build a better world." In the call to be "well composed," we sense the working of any number of forces that want to master us, to direct our gaze along prescribed trajectories, to subject our compositions to their disciplining view. Buñuel and Foucault seemed to us likely interlocutors here—Buñuel for his incisive questioning of normalcy through disturbing imagery and Foucault for his understanding of subjectivity as a product of panoptical disciplinarity. Both remind us of our interpolation as composed beings, offering us purchase on viewing the mastery of composition in process, of composed subjects coming both into being and into articulation. It worked. Viewers seemed provoked to have precisely the discussions we wanted to stimulate.

The second topic we encountered, and one that frequently preceded the discussion of the content of *Viewmaster*, was often rendered as a series of questions about our choice of genre itself, the installation: Why make an installation for a conference? Couldn't you have accomplished the same effects, provoked the same discussion, in a standard conference paper? The answer to the latter question is both yes and no; yes, we could have raised similar questions by giving a conference presentation, but no, we do not believe we could have induced the same sense of bodily urgency, the same materially felt necessity to consider these questions without the invitation into a separate space to experience a multidimensional, multimediated installation. After all, nearly everyone who experienced *Viewmaster* commented on the looming eyes that bore down constantly on visitors to our darkened room; those eyes provided a visceral reminder that questions about textual production, about the notion of the "composed" text, often go hand in hand with notions of the "composed" body, the disciplined "subject," the individual submitting work that falls inevitably under scrutiny, a gaze.

The former question—why make an installation at all?—raises issues about the nature and state of "composition" at this particular point in both the history of our discipline and the history of Western literacies. At a basic level, creating a multimedia installation for a conference on computers and writing seemed obvious. With the increasingly easy ability of computer users to create multimedia, and with writing instructors' increasing willingness to explore those multimedia dimensions of computer use with their students, experimenting with multimedia at the Computers and Writing Conference served as an exploration of what could be done with multimedia to provoke intellectual discussion. We had had some success with a previous installation (*Multimediated [E]Visceration*), at the 2008 Watson Conference on the "New Work of Composing," and we wanted in *Viewmaster* to continue the discussion about what exactly that new work of composing might be, particularly given the multimedia capabilities readily available through most computer and even mobile devices.

At the same time, *Viewmaster*—the ominous eyes, the specter of the panopticon, the call not just to compose but also to consider the nature of what is "well composed"—sounded what we hoped would be a questioning, perhaps even cautionary, note. For more than two decades now, our subfield of computers and composition studies has explored numerous technologies to enhance and expand the teaching of writing. We recognize that standard literacy practices have given way to what Stuart Selber has called "multiliteracies" (*Multiliteracies*), and that students increasingly need to be versed in a variety of textual, visual, and multimodal formats if they are to participate as literate citizens and workers in an increasingly multimediated world. This work has slowly trickled up, as it were, to the larger discipline of composition studies, as we can see from "CCCC Statement on the Multiple Uses of Writing" and its call to be cognizant, even respectful, of the diverse platforms through which people compose—and design—communication.

However, with that call for respect comes a series of expectations, not often articulated, about what constitutes a "composed" text, or the "right" kind of writing. As we undertake the "difficult work of helping students use good words, images, and other appropriate means, well composed, to build a better world," we pause to ask, what are *good* words? *Good* images? *Appropriate* means? Whose "better world" are we building? Some quick examples might illustrate why we feel the need to ask such questions.

TECHNO-COMP'S COLONIZING PEDAGOGIES

Following the lead of Selfe, many in the field of computers and composition bend over backwards to "pay attention" to student literacies, extracurricular compositional practices, and self-sponsored writing. Indeed, many of us in composition and writing studies have argued consistently for the increasing need to adopt new media technologies to help students develop media-rich texts for a variety of purposes, audiences, and critical aims. Daniel Anderson's hypertext/video essay, "Prosumer Approaches to New Media Composition," is a famous and oft-cited early example. Anderson argues that we need to teach students to produce new media, in part because our students already consume new media. Students who are "prosumers," Anderson writes, are likely to be more critical consumers. In *Multiliteracies for a Digital Age*, Selber usefully extends this argument in his call for the development not just of students' *literacy* but of their *multiliteracy*. Such multiliteracy inevitably entails training students to produce new media texts that demonstrate simultaneously a rhetorically savvy and critically engaged awareness of both the possibilities and the limits of new media and multimedia textual production.

Compositionists across the country promote such multiliteracy through a variety of courses and curricula. For instance, the Composition Program at the University of California, Irvine (UCI) is staffed by well-trained and professionally engaged compositionists who work hard to develop coursework that will help students develop more traditional academic literacies and think critically about the new communications platforms. The campus-mandated student learning outcomes for lower-division writing include these goals:

- demonstrate rhetorically effective, accurate writing and communication ability across a variety of contexts, purposes, audiences, and media using appropriate stance, genre, style, and organization;
- develop flexible strategies for generating, revising, editing, and proof-reading texts;
- develop abilities in critical reading across a variety of genres and media;
- demonstrate information literacy skills by locating, evaluating, and integrating information gathered from multiple sources into a research project. ("Writing Requirements")

The first (of two quarter-long) required composition course into which many UCI students place, WR39B, focuses on rhetorical awareness and analysis. Students read a range of texts, become familiar with the basic tenets of rhetorical analysis, and, at the end of the term, create their own rhetorically driven project. For several years, the final assignment, called "Rhetoric in Practice" (or RIP), has allowed the student to "determine the genre, purpose and rhetorical situation of [the] text, and then write/create the text." This process-driven assignment requires some attention to analysis, must address the class theme (which varies from section to section), and must "use texts read by the class as sources." The genre possibilities are often wide open, and student projects have varied from parodic commercial advertisements, to brochures for any number of (real and imagined) student services, to remixed sound files, to mashed-up comic books, to creative websites, to fictional Facebook profiles. Certainly, not all projects are equally creative, insightful, or polished. However, the point of the exercise is not to produce a polished "product," but rather to harness students' interest in multiple media and then prompt them to reflect on their rhetorical choices. In fact, the dominant grade for the project is often attached to a "memo" that students compose describing their projects, their choices, and their limitations. The goal is to see the extent to which students are sensitive to, and can begin actively and critically to participate in, basic audience-aware, rhetorical situations.

Faculty generally love this assignment, as do many students, and we could probably find its like in a variety of other large composition programs across the nation. These sorts of assignments have much to offer, particularly as they familiarize students with the functions and possibilities of rhetoric and genre while tapping into their multimodal interests. At the same time, however, this assignment has raised some questions that compositionists at UCI struggle to answer. For instance, in one RIP piece, which was voted "best" by an annual committee judging work produced by the Composition Program and submitted to a year-end contest, a student created a visually impressive animated film based on a short story by Haruki Murakami. The video animated part of the story about the anxious meditations of a young man facing a dead-end (and deadening) career. The video itself made little overt reference to the Murakami short story. As such, the audience was led to believe that the student had created an impressive animation about his own impending career anxieties. Certainly, the student may have shared the fears of Murakami's character. Within the context of the original class, the student was, perhaps not unjustly, rewarded for some wonderful technical mastery; and indeed, we should note that we are not criticizing the student's work *within the context of his class*, in which the student was probably encouraged by an instructor, who was encouraged by a curriculum, to repurpose material in such a multimediated way. However, to what extent is the video, however well done, simply an illustration of a text, as opposed to a thinking-through of the rhetorical capabilities of the video medium? Moreover, if the relationship between the original source material and the video illustration is unclear, how are we to understand—and how is the student to articulate *through the medium of video*—his relationship to different media?

In another case, Jonathan was called upon to serve as one of the annual judges in this category. He quickly discovered that not all RIPs are created equally. Not only did some students fail to reference source material appropriately, but a variety of different projects failed to engage any kind of substantive analytic. For instance, some RIPs consisted primarily of text rearranged artfully, or text bubbles in comics whited out and new conversation overlaid. At times, such mashups and revisions could rise to the level of parody. At other times, though, they were simply "mixed up." In one case, a duo of students took lines from a popular television show and redistributed them as tweets from individual characters on Twitter. While this project struck some faculty as particularly ingenious, other faculty were left puzzling over what had actually occurred in this project that was analytical or critical. Certainly, the students deployed some technical expertise—but to what end?

Some faculty have begun to wonder about the efficacy of allowing students to work with such material in the classroom. With so much else that must be taught to ensure that students have a chance of success in other courses and beyond, do we have time to incorporate into our classrooms a variety of new communications platforms currently available and teach them effectively? And when we leave the choice of platform, of delivery, open to our students, how can we ensure that students learn about the particular rhetorical affordances offered by the platforms? Certainly, the RIP projects—like many other assignments of the same ilk—are not designed to help students become proficient *producers* of new and multimedia texts. Pedagogically, they function to provide students opportunities to understand basic rhetorical strategies. At the same time, some faculty cannot help but wonder what messages are sent unintentionally by incorporating new media into the composition classroom in ways that do not fully allow students to explore the rhetorical capabilities of the media with which they work. In *Literacy in the New Media Age*, Gunther Kress opens his analysis of the rise of the visual field as a dominant literate domain in the age of new media by reminding us that "the two modes of writing and of image are each governed by distinct logics, and have distinctly different affordances" (1). If "writing" and "image" are indeed "governed by distinct logics" and indeed "have distinctly different affordances" (not to mention sound and video), then what do we lose—or, at worst, confuse—for students by perhaps too quickly adopting new media and multimedia projects into our classrooms, particularly when they serve to supplement other kinds of written assignments?

This last question vexes us increasingly as the broader discipline of composition studies encourages the use of new and multimedia in the composition classroom. We worry that the richness of multimedia, and the diversity of rhetorical affordances germane to the many dimensions of new and multimedia, may become lost as we incorporate these media into our classrooms and make them do the work of "writing." One final example should make our concern clear. Jonathan once posted on Facebook that he "should be writing" but was "making a video instead." The video, which he didn't describe, was just a ten-minute tour for his mother of his new house. Almost immediately a well-meaning and intelligent graduate student at a major PhD-granting program in our field wrote, "That IS writing." A brief conversation ensued, in which the student recognized, of course, that making a video is not really at all like writing, but we both were given pause by the initial sentiment, if only because it is one we have heard many times before. It's all writing. *Everything* is writing. Clearly, however, as we pointed out with Kress, not *everything* is writing; we would do well to remind ourselves that the "distinct logics" and "different affordances" of various media and modes are not reducible to one another. We ignore this fact to the detriment of our own and our students' understanding of the richness of new and multimedia.

PROPOSITIONS FOR (RE)IMAGINING TECHNO-COMP

The preceding examples and experiences have served in part as the inspiration for this book. To return to our installation, we intended *Viewmaster*, and we intend this book, to intervene at a particular moment—*this* moment—in the discipline's growing embrace of new and multimedia composition. As definitions of *literacy* expand, as conceptions of "text" transform, and as notions of composition pedagogy change to meet our students' needs, we want to take stock of how our field is changing and to chart its potential trajectories—for better and for worse. For while we have been among those who have at times eagerly embraced the possibilities and the transformations of "techno-comp," we sense the possibilities missing in that embrace, the left-out, the elided, the ignored that both misses the significant rhetorical affordances of new media and that, in our failure to teach those affordances, may hamper our students' use of new and multimedia.

On Multimodality hardly argues *against* using new media and multimedia in the teaching of writing. We believe—and assume a belief in our readership—that engaging compositional multimodality constitutes a pressingly necessary task for a wide variety of composition and writing studies courses. Increasingly, multimedia and multimodal composing have become key ways of meaning-making among younger generations of college students, and so developing a critically literate approach to such textual production seems crucial. We cannot disagree that exploring such rhetorical strategies forms a significant dimension of twenty-first-century literacy education. We have ourselves participated in the adoption of new media into the composition classroom, working with students on a variety of new media projects in addition to more formal, traditional academic essays. Our own writing classes—undergraduate and graduate—engage in the kinds of new media authoring, writing, and composing that our larger field envisions as possible and even desirable. If our classes work well, we explore the questions of possibility and desirability that underlie new media. Moreover, some scholars in composition studies are not just teaching but also producing scholarship in and through new media platforms. The journal *Kairos* has published digital scholarship for more than a decade, for example, and new ebook publishing venues are emerging, such as Utah State's Computers and Composition Digital Press and the University of Michigan's Sweetland Digital Rhetoric Collaborative. The excitement and energy around such ventures is palpable.

However, it is just such energy, such *felt* need to work with media, that gives us pause and has prompted us to write this book, asking for a greater pause for reflection on what precisely we are doing in US composition studies with multimedia. In our critical reflection on both our own practice and what we see as greater trends in our field, we would like to explore questions of discourse and ideology, drawing from Göran Therborn's definition of *ideology*. We would like to consider how Therborn's questions of "what is real," "what is possible," and "what is good" (18) play out in different iterations of technology, writing, rhetorical activity, and composing. More pointedly, we would like to trouble the waters of "good, real, and/or possible" in composition's intersections with new media. In so doing, we hope to get at questions of legitimacy, authority, and belonging. Specifically, we want to interrogate our engagement with—our attempted "adoption" of—new media's future.

With such questions in mind, we argue the following throughout *On Multimodality.*

First, we maintain that composition's embrace of new and multimedia often makes those media serve the rhetorical ends of writing and more print-based forms of composing. We are uneasy with the reduction of technology and *techne* to "skills" and "know-how," a reduction based on the emptying of new media of its excess, its generative power, just as we are uneasy with the reduction of writing to such things. While such reduction may necessarily be the case due to time constraints, we need to be aware—and apprise our students—of what we (and they) are missing.

Second, and stemming from the preceding critique, our embrace of new and multimedia for composing often ignores the unique rhetorical capabilities of different media, including the "distinct logics" and "different affordances" of those media. Put simply, we often elide such considerations—consciously or not—in order to colonize the production of multimedia texts with more print-driven compositional aims, biases, and predispositions. In the process, we hamper our students' appreciation of and ability to manipulate multimedia texts.

Third, and following from the preceding two critiques, we argue throughout this book that any approach to new and multimedia must become cognizant of the rich rhetorical capabilities of new media so that students' work with those media is enlivened, provoked, and made substantive. These media enable active engagement with a variety of complex public spheres. If students are to participate in increasingly technologized public spheres, they should be equipped to take full advantage of the specific rhetorical affordances of the media they are using. Most important, we need to remember that our invitation to students to compose multimodally and with multimedia is never innocent; as we invite the production of such "texts" into the composition classroom, we necessarily invite them to become composed in our eyes, under our watchful gazes. How we present multimedia and composing with multimedia will influence students' reception and production of multimedia. Shouldn't such reception and production be as robust as possible? Christian R. Weisser, in *Moving Beyond Academic Discourse: Composition Studies and the Public Sphere*, argues that

> since its emergence as an academic discipline in the 1960s, composition gradually expanded its focus from the individual writer, to social notions of how knowledge is produced, to more political investigations of discourse. I see the recent interest in public writing, public discourse, and public intellectualism as a continuation of that very expansion. More and more scholars in composition studies are interested in moving beyond academic discourse, in both the classroom and scholarship, and toward uses of discourse that might have more significance in shaping the world that we live in. (132)

If we want to make good on our efforts to teach students to participate productively in different public spheres, then we need to engage a more rhetorically sophisticated *techne* of such participation. Anything less, we contend, requires that composition do something else—that it eschew multimedia if it cannot teach it in ways that are fully cognizant of the rhetorical capabilities of those media.

A note on method: We proceed by asking how the rhetorical affordances of media might help us challenge ourselves to teach composing more robustly, with greater awareness of how to use different media effectively. We challenge ourselves and the field to think as critically as possible about the available means of persuasion at our disposal. To make good on that challenge, we frequently reference and pull from the long histories of media, since those histories offer us many examples of the particular rhetorical affordances of different media. Sheridan, Ridolfo, and Michel note in *The Available Means of Persuasion* that "multimodality is not new. Humans experience the world through multiple senses simultaneously, and practices of sociality (including rhetoric) have always reflected this" (xiv). Scholars in our field are increasingly paying attention to such histories. For instance, Adam J. Banks, writing in *Digital Griots: African American Rhetoric in a Multimedia Age*, describes the complex multimodal histories of DJing and how those histories have and might continue to inform African American rhetorical practice and intervention in multiple public spheres. Following Banks, we draw throughout this book on different histories of multimodal "practices of sociality" to highlight how we might teach students greater rhetorical awareness.

That said, we have not written a history of multimedia or multimodality. And our approach to history does not ignore the present; our move to historicize is to invite a more robust consideration of the multiple contexts—including the sociocultural, political, pedagogical, and affective—that inform, structure, and condition how we compose with new media. Such composing, we believe, is always "interested," invested in forwarding values and ideologies, consciously or not. Greater consciousness of how we compose with new media, as well as the horizons we see and can imagine for such composing, should invigorate our composing and pedagogies. At times, our interest in certain media forms, genres, modalities, and affordances may seem idiosyncratic, and it is; all of us have particular interests in media. As Banks draws on African American multimodal practices, we, as two queer compositionists, poets, and multimedia artists, draw on our knowledges of the histories of queer and avant-garde media to model for our readers one way to expand our sense of the rhetorical capabilities of media. In our pedagogical practice, we have found such avant-gardist media interventions to be among the richest for thinking about the rhetorical possibilities of different media. You will find others. Our aim remains to model a historical sensitivity while encouraging you to explore your own relationship to various histories of media (and there are many) with your students and in your own production of media-rich texts. Such historical sensitivity may be among the most important habits of mind we can offer our students for their own rhetorical education.

Selber justly notes the following:

> It is often claimed that computers have produced an enormous number of positive changes in higher education, changes that have vastly improved the social as well as instructional landscape that students and teachers inhabit. The trouble with such an unqualified claim is that it grants a level of autonomy to technology that simply does not exist. Although computers have the potential to assist the progress of positive change, they have just as much potential to help ensure the status quo. This is not to say that computers are neutral, but rather that teachers who are committed to progressive agenda for education must pay attention to far more than technology. (*Multiliteracies* 233)

We agree that computers have produced an "enormous number of positive changes," just as we question with Selber the seeming "autonomy" granted to technology. In particular, we appreciate Selber's caution that technology can assist in maintaining the status quo. Granted, if we as a field are satisfied with the status quo, then our arguments in this book need not be heeded. However, if we are invested in preparing our students for rich, critical engagement with media, then our sense of what composition does must change. Selber gestures toward such changes when he says, "I assume that departmental and institutional support systems will provide teachers with the theoretical background they will need in order to be alert to the social, political, and economic issues that are inextricably bound up with technology development and use" (*Multiliteracies* 233). To his list, we would also add the histories that such media are not only "inextricably bound up with" but out of which they arise and through which they continue to offer rich rhetorical possibilities.

LOOKING PAST "LEGITIMACY": ENGAGING MULTIMODALITY

In many ways, we may be arguing for a reconceptualization of composition studies that questions its most basic assumptions about what kinds of communicative practices should be privileged and what kinds of rhetorical affordances should be taught as we work with students to create rich, multimodal, and multimediated compositions. Following the work of Douglas Downs and Elizabeth Wardle in "Teaching about Writing, Righting Misconceptions: (Re)Envisioning 'First-Year Composition' as 'Introduction to Writing Studies,'" we hope to question productively how composition does its business and how it might do it differently. Downs and Wardle argue that

> if writing studies as a discipline is to have any authority over its own courses, our cornerstone course must resist conventional but inaccurate models of writing. A re-envisioned FYC shifts the central goal from teaching "academic writing" to *teaching realistic and useful conceptions of writing*—perhaps the most significant of which would be that writing is neither basic nor universal but content- and context-contingent and irreducibly complex. (557–58; authors' emphasis)

Our modification to this claim is that if composition as a discipline is to engage new and multimedia, it must "resist conventional but inaccurate models of writing" and think much more critically about new media as a variety of forms of situated communication possibilities. We must, more specifically, resist the universalizing desire to reduce all communication to simply "writing," but instead understand that new media, as a powerful possibility of communication, is "content—and context—contingent and irreducibly complex."

That contingent and irreducible complexity may be what most powerfully calls composition to reconsider its aims and goals—and perhaps even its weddedness to its own disciplinarity. Wikis, social media, new literacies, sexy new specialties, crossover potential, a chance to have "hidden knowledge"—all of these potential futures enliven our sense of ourselves as engaged compositionists serving our students' literacy interests and technological needs. At the same time, new media often becomes just another tool to make a claim about our own legitimacy; in itself, it is seen as little more than a set of skills, a way to manipulate particular programs, a familiarity with Prezi, HTML5, or Dreamweaver. To be sure, composition is deeply invested in its own legitimacy and in that of its texts (our scholarly journals as well as our students' texts)—and for good reason. We as a field are often charged with confirming students' abilities to "write" in "standard" English. Funding for programs relies on such confirming and assessing. Our own promotion and tenure depend on a similar legitimating move. How often must the free-form creativity of brainstorming and prewriting eventually give way to the composed text that presents its arguments in a complex (one hopes) but nonetheless rational, straightforward manner? Excess, often characterized as the extraneous, the "off-topic," must be trimmed to produce shapely texts that argue points, consider counterarguments, and reach reasonable, rational conclusions. New and multimedia, however, often (not always, granted, but often) play with excess, with the dis-composed, with possibilities of communication and rhetorical affects and effects that take us far beyond the reasonable, the rational, the composed. In this way, our work builds on that of Geoffrey Sirc's *English Composition as a Happening,* in which Sirc argues cogently that composition has, particularly in the last forty years, privileged the formal and composed text over other, messier, less-reasoned but nonetheless powerful forms of compositional practice. We take Sirc's basic idea and extend it to the multiple and diverse realms of new and multimedia. If we take new media seriously, we must be willing to risk the dis-composed text.

To flesh out such thoughts, we have divided this book into five chapters that view our argument from different and particular perspectives. The first chapter, "Refiguring Our Relationship to New Media," offers an overview of how composition studies has adopted new and multimedia. We use this chapter to look at some of the ideological and pedagogical pressures that have led to the widespread teaching of multimedia production in composition courses and how such production is framed primarily through the lens of "composing" and "writing," a lens that necessarily narrows students' view of new and multimedia as composed texts.

The next three chapters further this thesis by looking at particular genres of multimedia composing; doing so offers us a chance to meditate on how some composition instruction often colonizes multimedia with more text- and print-driven aims. In Chapter 2, for example ("Direct to Video: Rewriting the Literacy Narrative"), we examine how much video "authoring" in composition courses privileges a textual and linear narrative over other rhetorical possibilities offered throughout the history of video, particularly avant-gardist film. The third chapter, "Prosumerism, Photo Manipulation, and Queer Spectacle," rethinks photo manipulation as a popular invention technique in composition classrooms and argues for the power of images and mashups to create rhetorical possibilities that not only question standard narratives of gender, but also communicate powerfully in ways that a composed text cannot. The chapter further argues that we ourselves should engage in prosumerist multimodal writing, that we should, along with our students, critique and design multimodally if we are to participate in a changing public sphere. In the fourth chapter, "Collaboration, Interactivity, and the *Dérive* in Computer Gaming," we look at composition's recent love affair with computer gaming and note how many gaming practices actually challenge some of the sacred shibboleths of compositional practice, particularly around collaborative writing and the appropriate spheres for composing. We note in particular how many gamers practice the Debordian *dérive*, opening up new ways of conceiving mass market games by playing them against themselves and in ways far from the intentions of game designers. While games are certainly "useful" in the ways that James Paul Gee has suggested, their players often have ideas of their own about what to do with games, ideas that challenge us to rethink what to compose—and how—with new media technologies.

Finally, thinking critically about such provocation in the last chapter, "Theorizing the Multimodal Subject," we look at a variety of multimedia responses to the terrible shootings at Virginia Tech in 2007 and attempt to model a pedagogy that honors the diversity of multimedia production while turning a critical eye toward such production as well. As our students work more and more multimodally, examining such rhetorical dimensions allows us to see how the representation, if not even the construction, of subjectivity is changing. At stake then is an understanding of the historicity of subjectivity itself. Offering such a view, we contend, gestures toward some of the "irreducible complexity" of engaging and composing multimedia.

A caveat: we acknowledge fully that our explorations of multimodal and multimedia practices in these chapters come from our own experiences. We hope that our narration and analysis of those experiences thickens our theorizing about media, just as we use different theories to thicken our analyses of our pedagogical practices. Again, our reliance on avant-gardist media practices is one particular historical way in to thinking about multimodality's challenge to composition as a discipline. Many other possible paths lay about us, such as the use of non-Western rhetorical practices and their histories. We offer our path, not in an exhaustive or totalizing spirit, but as a nod toward alternate paths for conceiving rhetorical education, and as a way to model the necessity of undertaking such journeys. We are at times polemical, but ours is a polemic born out of deep pedagogical reflection. We hope to provoke your own reflection.

Further, a note about audience is appropriate here. We come to this project with a long history of work within both composition studies broadly and computers and writing in particular. We could not have arrived at our analyses without the inspired experimentation of many colleagues in computers and writing who push the boundaries of what composition studies knows and understands as its proper objects of study: writing, texts, composing, and the teaching of such. The names of our colleagues appear throughout this text, and many of our arguments grow out of their work and will seem familiar to them. We intend, then, the "you" of this text to be the field of composition studies more broadly—those in the field who come to the use of multimedia more recently, perhaps without the long investment of time in thinking about composing with multimedia that has characterized the subfield of computers and writing. Certainly, we do not claim to represent this subfield in its entirety; we know some within it who will challenge and even disagree with our analyses. But we offer this book as one path toward bridging composition studies and computers and writing. This is why we wanted to place the book in the CCCC Studies in

Writing and Rhetoric Series, as opposed to a series more squarely in computers and writing studies, or even with a digital press. And since print is still our profession's privileged medium, we have created a "traditional" book, even as we hope the book will challenge how you think about the affordances of all media. With that said, we urge you to read the book with your Web browser handy so that you can check our theorizing and polemic against the digital "texts" we reference. Your reading experience will only be enhanced if you read the book with your tech handy.

Ultimately, our book is a hopeful one, even as it critiques our field. Our hope is that with greater and more substantive attention to the media we teach, we cannot only enrich and enliven the field but also transform our curricula in ways that engage student learning and student writing all the more powerfully. We hope you will read it in that spirit.

1

Refiguring Our Relationship to New Media

OUR GOAL IN THIS BOOK IS TO THINK SYNERGISTICALLY about new media and composition studies, much as Kathleen Blake Yancey called for in her 2004 CCCC chair's address (printed later as "Made Not Only in Words: Composition in a New Key"). "Never before," she wrote nearly a decade ago, "has the proliferation of writings outside the academy so counterpointed the compositions inside. Never before have the technologies of writing contributed so quickly to the creation of new genres" (298). Yancey points to new curricula and expanded ways of thinking about what we do when we teach writing. At the same time as we look to the future, she asks us to look to the ancient canons of rhetoric:

> I would note that we have separated delivery and memory from invention, arrangement, and style in ways that are counterproductive. Let me further say that too often we treat them as discrete entities when in fact they are interrelated. . . . But as my options for delivering texts have widened—from the page to the screen to the networked screen and then back to the page anew—I've begun to see the canons not as discrete entities . . . but, rather, as related to each other in much the same way as the elements of Burke's pentad are related: the canons interact, and through that interaction they contribute to new exigencies for invention, arrangement, representation, and identity. Or: they change what is possible. (316–17)

Part of "what is possible" for Yancey, then, is a refiguring of how the canons might move in relation to one another, so that, for example, "a shift in the means of delivery [brings] invention and arrangement into a new relationship with each other" (317).

What role does technology play in these interrelationships? In a 2009 article published in *Computers and Composition*, James E. Porter summarizes the significance of technical knowledge to information retrieval, rhetorical crafting of compositions, and the subjects of "authoring" and "writing" in general:

> Technical knowledge about distribution options—i.e., how audiences are likely to access, engage, and interact with information—pertains in critical ways to rhetorical decisions about informational content, design, style, etc. In short, technical knowledge is integral to the art of rhetoric and to the canon of rhetorical delivery in the digital age. As Kathleen Welch (1990) argued nearly thirty years ago, "The fifth canon [delivery] . . . is now the most powerful canon of the five." Now more than ever. (208)

While we agree with Porter about the centrality of technology to the canon of delivery, we would like to think more broadly about the increasing centrality of technology, particularly the "new media," to the energetic interrelationship of canons of invention, arrangement, style, and perhaps even memory. A number of thinkers in our field, from established scholars such as Yancey to emerging techno-theorists such as Collin Gifford Brooke and Ben McCorkle, have proposed that our understanding of rhetorical canons must be reconfigured, or at least made less static, given increasingly technologized delivery of content. Indeed, such is the focus of a recent volume, Stuart A. Selber's 2010 collection *Rhetoric and Technologies: New Directions in Writing and Communication*. In his introduction, Selber writes:

> [It] is difficult to imagine a rhetorical activity untouched by ongoing developments in writing and communication technologies. Their increasingly widespread integration into all facets of culture has encouraged scholars and teachers to reinterpret (yet again) the traditional canons of rhetoric. Invention strategies, for instance, now address powerful search capabilities and the ways in which database structures shape access to an intellectual landscape. Rhetorical education on

> arrangement no longer assumes a linear organizational pattern—or a patient reader, for that matter. More than occasionally, writers and communicators today anticipate reader control with modular hypertexts that can support multiple interpretive pathways and that can invite textual transformations and revisions. (2)

At this point, announcing the inter-imbrication of technology and rhetoricity seems almost a truism in our field. Yet truisms often contain within them untested, even unacknowledged assumptions. Obviously, many specialists in computers and writing are increasingly aware of the histories of media, and as this interdisciplinary subfield grows, its engagement with the complex histories of media and technologies informs its understanding of composing with and in digital and technologized environments. However, we would like to explore the troubled contact zone between the subfield and the larger field of composition studies—a field that often figures new media and multimedia as somehow subservient to the interests of a more traditional, essayistic, and print-driven conception of composition.

This chapter traces the emergence and use of new media technologies in the field of composition studies. We note that our field has attempted to "pay attention" and take seriously the impact of new media and other new communications technologies on our understanding of composition in general. At the same time, we argue that this engagement with new media is often structured through (1) an emphasis on the rhetorical capabilities of textuality and (2) a concurrent elision of the rhetorical affordances of multimodality and multimedia.

A BRIEF HISTORY OF COMPOSITION AND NEW MEDIA, OR, WHAT DO WE TALK ABOUT WHEN WE TALK ABOUT TECHNOLOGY?

A Sense of Threat . . .

While we can hardly provide an exhaustive history, we can catch a strong sense of our field's long and complicated engagement with computer technologies in the scholarship presented in the international journal *Computers and Composition*, well into its third decade of publication at the time of this writing. At the same time, taking a broad and high-level view of major publications in the field—publications that continue to be cited and have exerted a significant influence on our understanding of writing technologies and new media—offers us a telling set of insights into how our field has understood, incorporated, and in some ways attempted to colonize new media.

While some in our field labored nearly three decades ago to think critically about the use of computers in the composition classroom, much of that work focused on word processing and the digital generation of text; as Charles Moran notes, it was a time during which we hoped for "the simple efficacy of technology for teachers and for writers" (352). With the advent of the World Wide Web in the mid-1990s, many compositionists announced the emergence of a new era in our understanding of textuality, literate practice, and compositional possibility. As Moran points out in his survey of *Computers and Composition* scholarship, 1996–97 was a "watershed" moment in our thinking about computers and literacy; in particular, that time saw a new focus on the connections between writing technologies and literacy pedagogies (352). This sense that technology and literacy would renew each other finds an echo in Jonathan's 2006 *Digital Youth: Emerging Literacies on the World Wide Web*, with the rather portentous opening statement that "literacy is changing" (1). Many at the time felt—as many continue to feel—that literacy is indeed changing, that new media technologies, represented most powerfully by the Web, signaled a profound change not only in terms of the speed or the access with which individuals retrieve information, but also increasingly with the forms, modalities, and genres through which people create content, engage others, and disseminate a diversity of views. Jonathan, among others, recounts the early debates in the public sphere—both pro and con—about the potential effects—both pro and con—of increasingly widespread engagement with and through the new communications technologies on a diversity of issues besides literacy, including possible effects on children's safety, the archiving (or disappearance) of information, surveillance, the public sphere, and the state of democracy itself.

This book touches variously on these and other important concerns. But if we pull back for a moment to examine how our field, as a disciplinary construct, acknowledged and understood the importance of the new media technologies, we see a startling shift. For instance, Bloom, Daiker, and White's 1996 collection, *Composition in the Twenty-First Century: Crisis and Change*, contained only one chapter about computers, technology, and writing: Andrea A. Lunsford's thoughtful "Intellectual Property in an Age of Information: What Is at Stake for Composition Studies?" For a volume that announces itself as inviting thinking about composition in the twenty-first century, its nearly complete lack of engagement with issues of technology and new media seems, in retrospect, a profound oversight. But we must remember that a book published in 1996 would have been in the works for a couple of years prior, so the contributors are writing just before the 1995 turning point, as it were, in thinking about technology and literacy. For indeed, by 2003 the editors' new volume, *Composition Studies in the New Millennium: Rereading the Past, Rewriting the Future*, contains an entire section on computers and writing with the rather foreboding title "How Will New Technologies Change Composition Studies?" The implicit threat—that new technologies of writing will somehow change everything about composition—comes through at times even in chapters that do not focus on computers and writing. For instance, early in the collection, Susan Miller worries that "we have increasing difficulty in maintaining the ethos that knows better than students the appropriate uses of writing and that defines them as ancillary to personal, not social, development"—and this statement comes in an article entitled "Why Composition Studies Disappeared and What Happened Then" (55).

This tone of threat is worth keeping in mind, if only because it seems to have been overcome and in some ways forgotten in the last decade. Certainly, not everyone in our field embraces new media technologies, but the majority seem to hold their arms open to using them in the composition classroom, even if only as information retrieval. Many others, though, have understood that the new technologies have indeed changed composition studies, so much so that our profession's national organization, the Conference on College Composition and Communication, has issued at least three statements that deal substantively with "electronic issues"—including the statements "Teaching, Learning, and Assessing Writing in Digital Environments" (2004), "Principles and Practices in Electronic Portfolios" (2007), and "Statement on the Multiple Uses of Writing" (2007). Our sense of threat has turned into active engagement.

But what kind of engagement? Where has the sense of threat gone? Has it actually left our field, or might it have been transmuted into other impulses, perhaps a desire to contain and control that threat?

Threats . . . and Promises

We can catch a glimpse of possible answers to such questions by attending to the development and evolution of the discussion of what to do with the new technologies. DeVoss, Johansen, Selfe, and Williams's contribution to Bloom, Daiker, and White's 2003 collection is titled "Under the Radar of Composition Programs: Glimpsing the Future through Case Studies of Literacy in Electronic Contexts," and the authors' opening salvo makes the changing situation for composition and writing studies quite clear by posing a leading question:

> What new understandings of terms such as text and composing will students bring with them to the college classroom in the next decade—especially those students habituated to reading and composing the kinds of new-media texts that have come to characterize contemporary computer-based environments? (157)

Our students, flying at the time "under the radar of composition programs," will inevitably force change in our understanding of "text and composing" as they bring their own understandings of these concepts to the college classroom. DeVoss and colleagues attempt to show us both the possibilities and the potential benefits of embracing such changes, but the sense of threat lurks in the militaristic imagery of the title: the kids are sneaking up on us, and we'd best be prepared. At the very least, we have the chance to stay a step ahead.

To be fair, DeVoss and colleagues offer in their chapter some key insights, not just about the new communications technologies, but about how we should approach and understand any literacy practice. For instance, in their concluding "lessons" about what the new technologies have to teach us about literacy and writing, they note:

> Lesson 1. Literacies have life spans linked to the cultural ecology of a specific time and place. Depending on a complex of circumstances, literacies emerge, accumulate, compete, and fade at varying rates. English-composition teachers and programs need to acknowledge, understand, and respond to this dynamic. (168)

This important insight acknowledges that literacy practices find their roots in historical moments, in "specific time[s] and place[s]." Earlier, Selfe presages this notion with coeditor Gail Hawisher, writing in their introduction to *Global Literacies and the World-Wide Web* (Hawisher and Selfe, 2000) that their book examines

> culturally specific literacy practices—authoring, designing, reading, analyzing, interpreting—on the Web, especially as these practices are shaped, both directly and indirectly, by concrete contexts for language and language use. The overarching goal of the volume is to test the commonly accepted premise that the Web provides individuals around the globe with a common and neutral literacy environment within which international communications are authored, read, and exchanged. (3)

Hawisher and Selfe, like many in the field of computers and composition studies, know that the Web does not constitute a neutral compositional space, and that people who compose for the Web, who use new and multimedia, work in specific sociocultural contexts, bounded by intricacies of location, access, ability, and ideology. Since those contexts change over time, any understanding of literate practice must acknowledge those changes, tracking their histories and developments to achieve a richer sense of the horizons—and limits—of literate practice and possibility.

However, while this insight into the historicity of compositional and literate practice accompanies the field's serious turn toward thinking critically about new media platforms as compositional spaces, other competing insights emerge as well. And curiously, competition seems somehow central, if barely acknowledged, as a shaping force in discussions of emerging "techno-composition." DeVoss and colleagues actually acknowledge a sense of competition in describing the varied "landscape" of literate practice:

> Increasingly, the literacies practiced by individuals who communicate primarily in online environments exist within a dynamic cultural ecology influenced by expanding global markets and computer networks that stretch across language barriers, cultural groupings, and geopolitical borders. Within this ecology, as the New London Group and [Gunther] Kress have explained, new-media literacies—which rely as much on images, video clips, animation, sound, and still-photography as on words—have begun to emerge and compete vigorously with more traditional alphabetic print texts for readers' attention. (168)

The competition between "new-media literacies" and "more traditional alphabetic print texts" suggests ramifications for composition and writing curricula, resulting in the authors' positing of "Lesson 2":

> English-composition teachers and programs must be willing to address an increasingly broad range of literacies—emerging, competing, and fading—if they want their instructions to remain relevant to students' changing communication needs and experiences within the contemporary cultural ecology. (169)

The lesson is simple. Our relevance as composition instructors is at stake. Hence, we must pay attention to the "contemporary cultural ecology" and find a way to "address an increasingly broad range of literacies." The third lesson the authors propose suggests that "English-composition teachers and programs need to start with the literacies that students bring to the table and in which students are invested, but they can't stop there" (170).

The inevitable question: where do we go?

Round and Round We Go . . .

Lester Faigley's essay in the section "How Will New Technologies Change Composition Studies?" (Bloom, Daiker, and White, *Composition Studies*) offers a complex answer to that question. In "The Challenge of the Multimedia Essay," Faigley points out that we are broaching a disciplinary divide—namely, the divide between composition as a text-based discipline on the one hand and communication studies and technical writing as a somewhat broader endeavor on the other:

> Much scholarship in composition studies during the disciplinary period has argued for a place in the university beyond the limited conception of college writing that dominated for most of the past century. Yet, an attitude that Diana George and John Trimbur describe as a longstanding "counterposition of composition and communication" results in a polarization of the verbal and the visual. This polarization denies the materiality of writing as one means of visual communication alongside others, and it follows the bias of the humanities to distrust images not considered high art. Until quite recently, the field has enforced this distinction in writing courses by confining instruction in design, images, and graphics to technical writing. A neat and unproblematic compartmentalization of genre, however, is no longer possible. (186)

We agree—and more so. What about film studies, informatics, and computer science? Increasingly, we in composition studies more broadly poach on their "territories," suggesting that the diverse cultural ecologies of contemporary literate practice call our artificially divided disciplines into question. Our various fields and subfields seem to collide with one another, resulting in a questioning, as Faigley has it, of "genre." What are the contemporary genres of communication, composition, and delivery?

In a curious move, Faigley refrains from advocating what he seems to move toward: a dissolution of disciplinary boundaries. Rather, he sees the potential solution to the problem in strikingly simple terms:

> The point is not that we have to abandon teaching the traditional essay in writing classes. I am confident that essays will continue to be written long into the future just as short stories are today, even though the readership has vanished. However, we do not have to restrict our definition of the essay to what we have come to know as school form. (187)

Put another way, we can have our cake and eat it too. We can continue to teach the standard composed essay, but we can also broaden our understanding of what the essay is to accommodate a variety of forms, media modalities, and emerging literacy practices. In other words, all of this worry over "How Will New Technologies Change Composition Studies?" can be resolved if we think of everything as an "essay."

Granted, Faigley is one author in one collection, and this is one moment in the field's evolving engagement with the new communications technologies. But interestingly enough, precisely this notion of the "essay" lies at the heart of other debates in the field leading up to this exchange in Bloom, Daiker, and White's 2003 collection. Throughout the late 1990s and early 2000s, an increased interest in alternative rhetorics, discourses, and genres ran parallel to the emerging focus on new media and the use of communications technologies in the teaching of writing—at times overlapping one another. Bloom and colleagues' two collections staked important claims for thinking beyond the traditional academic essay. In 2001, editors Laura Gray-Rosendale and Sibylle Gruber proposed in the introduction to their collection, *Alternative Rhetorics: Challenges to the Rhetorical Tradition*, that a turn toward fostering alternative rhetorics would assist greatly in the critical pedagogical enterprise by questioning received wisdom about writing and its critical power, as well as traditional and hierarchical approaches often embedded in composed forms. Specifically, they argue that

> "alternative" proposes rhetorical approaches that draw from as well as disrupt and challenge the hierarchical nature of some traditional rhetorical studies while recognizing that such challenges are temporary and open to co-optation. . . . "Alternative" also testifies to the power relations involved in offering up these texts. These are texts which canvas rhetorics that have often been marginalized, ghettoized, neglected, or overlooked within our historical context as well as other historical contexts for particular cultural, social, and political reasons. (4)

On the one hand, Gray-Rosendale and Gruber propose something of a "recovery" project, attempting to make space for stories, traditions, and rhetorical practices that have traditionally been excluded from composition curricula. On the other hand, they begin to theorize how "traditional" formats, such as the composed essay, may carry ideological predispositions and values, however implicit or unconscious. For instance, the traditional essay's reliance on the movement from claim to claim, supported through logical argument, may very well overlook claims, positions, or insights grounded in emotion, or embodied experiences that might seem illogical or unreasonable when translated into syllogistically driven arguments.

Just a year later another collection picked up on precisely these notions and even more radically called for greater discourse and genre awareness within composition studies. In 2002's *ALT DIS: Alternative Discourses and the Academy*, editors Christopher Schroeder, Helen Fox, and Patricia Bizzell bring together a variety of authors to consider the value of alternative, sometimes non-Western rhetorical traditions that do not rely on Western or Aristotelian logic. For Schroeder in particular, two compelling issues are at stake. First, in his *ReInventing the University: Literacies and Legitimacy in the Postmodern Academy*, Schroeder recognizes our field's competing attention to the need to "certify" students capable of certain kinds of academic writing skills and the desire to pay attention to and honor students' diverse, nonstandard literacy practices. He notes the work of many in composition studies to enhance and enlarge our sense of literate practice, calling them "advocates for the legitimacy of students' primary discourses in classrooms conventionally dominated exclusively by academic literacies" (19). In this way, Schroeder picks up on a significant notion of the emerging techno-compositionists: the importance of paying attention to student literacy practices. Moreover, like Gray-Rosendale and Gruber, he sees a strong connection between literate practice and cultural value, arguing that "the act of becoming literate amounts to being certified in certain social practices and, through them, in particular cultural values" (5). Schroeder wants to make a place for understanding the critical efficacy of alternative rhetorics and discourses in composition classrooms, and he sees narrative as a powerful genre to forward this goal. He even argues, provocatively, "that (suppressed) personal narratives are central to intellectual work" (30). Indeed, "narrative," broadly conceived, is often profoundly rhetorical, even as many narratives—narratives of witness, in particular—do not rely on logical or syllogistic thinking.

As noted, a set of intense considerations about technologized literacy and rhetorical practices runs parallel to such discussions in the broader field, considerations of student access to and experimentation with a variety of nontraditional forms, genres, and modalities. A pivotal text, Hawisher and Selfe's 1999 collection, *Passions, Pedagogies, and 21st Century Technologies*, attempts to map the impact of technology and digitality on both student literacy practices and the implications of those practices for our field. The collection starts with a tension comparable to the one Schroeder identifies—the need of our field to "certify" students as proficient in certain kinds of literate practice and the desire to embrace a broader understanding of literacy. For instance, Dennis Baron, in the first chapter, argues that "my contention in this essay is a modest one: the computer is simply the latest step in a long line of writing technologies" (17); hence, no need for grand pronouncements about massive reconceptualizations of literacy. Similarly, in an essay early in the collection, Doug Hesse worries about "saving a place for essayistic literacy," so the collection quickly sounds a note of mild alarm: let's not lose what we have, and (curiously) there may be nothing to worry about anyway. At the same time, however, Myka Vielstimmig (Kathleen Blake Yancey and Michael Spooner) create a hybrid "essay," in which s/he maintains:

> This is not an argument against The Essay or against "print classic" or conventional logic. It is an argument toward another kind of essay: a text that accommodates narrative and exposition and pattern, all three. It allows for differentiation without exclusion, such that it resists becoming unified in a community of shared final ends, to borrow from Susan Miller. It is an essay of radically different identity politics, of radically different mentality. (91)

The alternative nature of Vielstimmig's essay gestures toward the larger debates in the field about the place of alternative rhetorics and discourses, creating a noticeable synchronicity with Schroeder's pronouncements on the importance of recovering the "(suppressed) narrative" as a powerful rhetorical form. At stake here is an entire politics of writing, and a vision for writing, particularly technologized writing, as potential political action. The disjointed and imagistic form of Vielstimmig's essay suggests that arguments, debates, and insights can be built in "essays" that stretch, perhaps even break, the boundaries of traditional forms. And technological platforms, with their ability to facilitate the use of images, sounds, and video, increasingly problematize traditional generic conventions.

Rereading these pieces at the distance of a decade can still give one a thrill; the thinking—and composing—is heady, inducing a feeling that you, with the authors, look over a precipice into a great, intriguing unknown. At precisely this heady moment, two other distinct voices entered to problematize the discussion, voices that question some of our discipline's basic assumptions and that critique its general lack of historicity.

First, Hawisher and Selfe's collection (*Passions*) includes Anne Frances Wysocki and Johndan Johnson-Eilola's important essay, "Blinded by the Letter: Why Are We Using Literacy as a Metaphor for Everything Else?" In terms of form, this "essay" is much like Vielstimmig's in that it plays with the genre of the logically flowing and developing essay, including pictures, proceeding as a set of interchanges, and interrupting itself to digress. It turns more on key questions than on offering supportable claims—but its questions are devastating. The authors ask us to consider why we view multimedia and multimodal compositions through the lens of literacy, particularly since these multimedia and multimodal formats engage visual and aural practices that traditional understandings of literacy, with their reliance on the alphanumeric, may not be able to encompass. Their questions echo those of scholars who advocate for exploration of alternative rhetorical practices in that they worry over what predispositions or assumptions about composing we may bring to multimodal composition as we view such composition as a form of "literate" practice. They ask, for instance, "what are we likely to carry with us when we ask that our relationship with all technologies should be like that we have with the technology of printed words?" (349), and "when we speak of the relationship we hope to establish—for ourselves and for our students—with newer technologies, do we want to carry forward all these particular attachments and meanings and possibilities?" (360).

With such questions, Wysocki and Johnson-Eilola suggest that we must think about the nontextual dimensions of multimedia composing in ways that do not see them just as tools to advance the kind of thinking associated with print texts. They argue powerfully that visual and aural dimensions of multimedia composition may very well have rhetorical affordances that exceed those of the "technology of the printed word," and they ask, "what other possibilities might we use for expressing our relationships with and within technologies?" (349). Like Vielstimmig, Wysocki and Johnson-Eilola set the stakes high by tying traditional, print-based literacy practice to particular ways of knowing and thinking; their language is intense and provocative:

> If "literacy" is already closely tied to our sense of how the world was colonized and settled and tamed, if "literacy" is already (deceptively) tied to political and social and economic improvement, if "literacy" already is the boundary of our sense of who we are, then why not apply the notion to newer technologies? (360)

The answer is, obviously, we shouldn't apply older notions of literacy to the new technologies. We should instead explore what those newer technologies offer us by way of an expanded notion of rhetorical practice and engagement with different discourses. Wysocki and Johnson-Eilola refer to literacy practices as "colonizing," bringing into different domains a set of predispositions and values—such as the preeminence of logic over other kinds of thinking. These authors thus open up a space for thinking about how we, as compositionists, might be colonizing new rhetorical domains with our own particular kind of thinking, our own predispositions and privileging of particular forms, such as the composed essay. Wysocki and Johnson-Eilola do not offer much more than such suggestive questioning, but we mark this oft-cited and highly influential essay as a pivotal moment in opening a space for thinking critically about our discipline's preoccupation with print-based literacies.

Along these same lines, Hawisher and Selfe's collection also contains Geoffrey Sirc's essay "'What Is Composition . . . ?' After Duchamp (Notes toward a General Teleintertext)," which would become a chapter in Sirc's important and influential book *English Composition as a Happening*. Sirc's appearance in this collection, we argue, is as significant as Wysocki and Johnson-Eilola's in that Sirc is deeply invested in exploring the spaces outside the immediate and bounded domains of composition. In particular, Sirc offers the beginnings of a necessary historical turn in discussions about technology's impact on composition; specifically, he argues—and fleshes out this argument brilliantly in *English Composition as a Happening*—that composition studies made a concerted effort in the 1960s to delimit its understanding of writing to the production of the "composed" essay. In the process, the field began largely ignoring avant-gardist experimentations in writing, which often pushed the boundaries of composed forms to the breaking point in an effort to provoke critical thinking about traditional forms and their attendant values.

Such writing—often experimenting with different media, hence the connection to technology and new media—was provocative, and frequently embodied and nonrational. It played with alternative discourses as a way to attempt to know differently. Composition as a field eschewed such play, according to Sirc, to put more attention into enabling students to create composed, academic texts that would, in Schroeder's word, "certify" them for participation in academic discourse communities. As such, David Bartholomae's "Inventing the University," with its call for paying attention to the process through which students first mimic and then "own" academic and discipline-based writing, seems to Sirc to embody a particular, and highly delimited, trajectory for teaching students about the creative and critical possibilities of writing. Sirc argues that for Bartholomae, composition "never explores the possible, just possible versions of the preferred" ("What Is Composition" 194). Sirc shows us how the "essay" wins out, and he is one of the few scholars at the time calling for a reconsideration of our fetishization of the composed ("perfected, conventionalized, ritualized") essay (194) and for the possibilities opened up by thinking about past histories of writing and compositional practice. As he notes later, his purpose is "to retrace the road not taken in Composition Studies, to re-read the elision, in order to remember what was missed and to salvage what can still be recovered" (*English* 13).

Ultimately, *Passions, Pedagogies, and 21st Century Technologies* set the stage for understanding the debates and challenges arising out of composition studies' embrace of both technology and multimedia/multimodal forms of composing. The field at the time seemed poised to embrace multimedia composing, particularly as a powerful form of alternative rhetorical practice. At the same time, however, many either worried over the loss of the "essay" or, more curiously, like Faigley a few years later, began to see multimedia composing through the lens of the essay. Others offer alternative ways to frame the relationship between technology and composition; most notably, Wysocki and Johnson-Eilola challenge us to rethink our disciplinary reliance on literacy and Sirc calls for greater attention to histories of compositional practice. But such calls seem to remain at the margins.

Even a very recent collection seems to hedge its bets a bit, wanting to encourage multimodal composition but worrying nonetheless over more traditional literacies. In *Multimodal Literacies and Emerging Genres* (2013), editors Tracy Bowen and Carl Whithaus note what, for many of us, has become nearly a commonplace in our thinking about authoring, composing, and writing in our composition courses and programs: "Within college writing courses, the emergence of a wide array of information and communication technologies (ICTs) in the last twenty years has opened up new possibilities for the types of compositions that students can create" (1). This edited collection seeks to "demonstrate how faculty and students are already exploring 'what else is possible' in these new media writing spaces" (1), and the authors focus in particular on issues of genre and how "multimodal student writing is doing something new—it's reshaping genre boundaries and changing what counts as academic knowledge" (4). Thus, Bowen and Whithaus maintain that "our job is neither to lead them into this changing world of multimodality nor is it to hold them back from it. Rather, we are in the midst of a shift that is affecting how we write, why we write, and where we write . . . or don't" (5).

We wonder why the editors hold back from pushing forward, from embracing new media and multimodal composition. They suggest that they neither want to "lead" nor "hold" students back, perhaps acknowledging that students come to our writing courses knowing (potentially) much more than we do, at least technically, about new media platforms. Just a bit further in their introduction, however, we begin to perceive some potential pressures that work against a full embrace of technologized compositional platforms. Bowen and Whithaus note that the act of encouraging multimodal compositions itself creates spaces within which new genres emerge, whether formally named or not (6). Yet they hesitate: "For writing program administrators, multimodal literacies bring new challenges—faculty and students need to explore the potentials of multimodal composing without losing the programmatic structures which facilitate the development of discrete writing skills" (9). Bowen and Whithaus want to encourage multimodal literacy, and their collection offers numerous superb examples of how to do so. However, those literacies must somehow fit within existing "programmatic structures" that are, admittedly, devoted to the "development of discrete writing skills." What are those skills? Bowen and Whithaus do not spell them out, but we can certainly discern here a tension between the new and emerging genres and the more traditional (one can assume essayistic) genres. Perhaps Bowen and Whithaus, writing from relatively new stand-alone writing programs, are highly sensitive to the potential fragility of the recently constructed disciplinarity of composition studies—a disciplinarity threatened by genres that do not yet even have names. After all, what is a discipline without discrete objects of study? How do we accommodate such looseness while positing our own discrete borders? Again, new media and multimodality—even in the friendliest hands—can seem like a potential threat.

Given this background, we argue in the next section that the field largely fluctuates between (1) the instruction of traditional alphabetic literacies and the creation of the "composed" text and (2) an attempt to reconfigure the "essay" through multimedia and multimodality. Throughout, the creation of the "essay" or the composed text remains dominant, regardless of the medium. And while we do not want to deride the importance of either the essay or the notion of the composed text, we note in the following pages that the insistence on seeing new media and multimodality through the lens of the "essay" is not without certain effects—effects that delimit our understanding of what and how new media move rhetorically.

THE WILL TO NORM: PEDAGOGIES OF TECHNO-INCLUSIONISM

One of the advantages of the emergence of "techno-comp," as we might call it, is the recognition that new media, like any set of "texts," are always consumed within particular social, cultural, historical, and even political contexts. Many techno-compositionists, as we saw with DeVoss et al., reference the New London Group and the work of New Literacy scholars such as Brian Street in situating their analyses of multimedia and multimodal textualities and compositional practices. In many ways, Wysocki and Johnson-Eilola (and to a lesser extent Sirc) acknowledge this New Literacy turn and bring to the table a historicized understanding of new media literacy practices. They contend, to varying degrees, that a historicized approach to new media allows us both an expanded notion of compositional practice and the capacity to make more informed curricular choices.

For better or for worse, this supple view often gets lost in the rush toward what we might call "techno-inclusionism," or the drive to include the new information and communication technologies in the composing process and in our curricula. The result is a colonizing (to use Wysocki and Johnson-Eilola's term) or a co-optation (to borrow from Gray-Rosendale and Gruber) of technological platforms for the promulgation of traditional essayistic literacies.

The will to norm technologies, as it were, has at times subtle and not-so-subtle manifestations. Even those who have been at the forefront of paying attention to and thinking critically about computers and composition have, at times, been ambivalent about what precisely to do with the new technologies. As we have seen, for example, Hawisher and Selfe (*Passions*) argued powerfully for paying attention to student literacies, or to what Deborah Brandt would call "self-sponsored" literacies, particularly as they engage technological platforms. However, in Robert P. Yagelski and Scott A. Leonard's 2002 collection, *The Relevance of English: Teaching That Matters in Students' Lives*, Richard J. Selfe and Cynthia L. Selfe discuss the relationship between critical technological literacy and English studies, arguing ultimately that

> our obligation as humanists and teachers suggest[s] that we must continue to learn with students about our humanistic responsibilities in a technological age, continue to insist on developing informed and critical perspectives on technology and technological literacy issues, and continue to insist on

> our roles as social agents in determining appropriate and productive uses of technology. (377)

Of particular interest in this list of imperatives is the desire both to "learn with students" and to "insist on our roles . . . in determining appropriate and productive uses of technology." The earlier "threat" of technology that played in the background of DeVoss et al.'s work is met in this contemporaneous piece by one of DeVoss's coauthors with an assertion of control. We must guide and direct students' use of technology, determining its "appropriate and productive" use. Curiously, this phrasing will become instantiated in "CCCC Statement on the Multiple Uses of Writing," with which we opened our introduction:

> The CCCC hereby calls together—and calls to action—all those who share its vision of a future in which an expansive writing curriculum, backed by ample resources, attends unyieldingly to the difficult work of helping students use good words, images, and other appropriate means, well composed, to build a better world.

Our embrace of these technologies, then, is an embrace to compose them, make them appropriate and productive, have them do the work of building our "better world."

Such generalized statements, though, obviously beg the questions: what is "appropriate" and "productive," and what constitutes the composition of a "better world"?

The answers to these questions, we argue, riddle the various sourcebooks, guides, and statements generated by the profession simultaneously to promote "techno-inclusionism" but also to control it, to compose it so that essayistic literacy practices are foregrounded. For instance, the Bedford/St.Martin's "sourcebooks" have served as important collections of seminal essays and articles addressing pressing professional and pedagogical issues. They also serve a disciplining function in that they shape the emerging canon of work in composition and writing studies more broadly, thus promulgating particular values and ideas as part of the discipline's foundational ethos. The emergence in 2008 of *Computers in the*

Composition Classroom: A Critical Sourcebook, edited by Michelle Sidler, Richard Morris, and Elizabeth Overman Smith, signaled the increasing significance for the discipline at large of the importance of the new communications technologies. While the selection of articles represents a range of thinking and viewpoints, the introduction guides readers in how to think about computers in the classroom, and the editors' characterization positions computers as aids to the composing process; note this summary, for instance, of the section "Writers and Composing":

> Writers and Composing presents a series of practical strategies for implementing computer technologies in the composition classroom, focusing on major components of traditional composition courses, such as research, drafting, workshopping, and proofreading. In addition, the readings explore how computer technologies alter the role of the teacher from one of evaluator to one of guide, a threatening transition for some new teachers. All of the writers emphasize that implementing technology does not simply add a computer tool to the classroom—it changes the writing, learning, and teaching environment in which we conduct composition instruction. Moreover, technologies alter students' literate practices, mediating how they compose, communicate, and research. (6)

Certainly, this passage gestures toward necessary changes brought on by techno-inclusionism, and the last sentence offers a token acknowledgment of "students' [changing] literacy practices," but the guiding emphasis remains on computers in relation to the "major components of traditional composition courses."

We find a similar emphasis on compositional knowledge in *Visual Rhetoric in a Digital World: A Critical Sourcebook*, edited by Carolyn Handa. In her introduction, Handa emphasizes what the composition instructor can offer the analysis of visual "texts":

> Students who possess a high degree of technological skill may see the value in knowing how to create a document using the latest digital tool but not understand the importance of think-

> ing carefully about rhetorical questions such as the appropriate audience, purpose, tone, and argument (Shauf). Here is where our experience as composition teachers becomes crucial. Preparing students to communicate in the digital world using a full range of rhetorical skills will enable them to analyze and critique both the technological tools and the multimodal texts produced with those tools. Visual rhetoric in the composition course then serves two ends: to help students better understand how images persuade on their own terms and in the context of multimodal texts, and to help students make more rhetorically informed decisions as they compose visual genres. (3)

To be sure, Handa's goal, like that of the editors of *Computers in the Composition Classroom: A Critical Sourcebook*, is to introduce a range of instructors of often very different standing within the university (and not all with professional backgrounds in composition studies) to pressing concerns and trends in the field—hence her useful insistence that composition teachers' experience can be "crucial" in helping students navigate the visual field, particularly given the prominence of the visual field in multimodal texts. And to be fair, Handa's anthology contains essays and articles from a range of people outside composition studies, such as Roland Barthes, graphic designer Jeffery Keedy, and the cartoonist Scott McCloud. Indeed, Handa, like the other editors, gestures toward greater concerns, such as how "images persuade on their own terms." But in the effort to make sure that composition instructors are comfortable with this new media, all of these editors foreground what compositionists already know, already do, already value—skills and strategies developed through writing process pedagogies, attention to the kinds of rhetorical questions we ask about persuasive texts, and, overall, the role of the composition instructor as guide, mentor, and "social agent," to borrow from Selfe and Selfe.

This focus, however understandable, fosters a techno-inclusionism that always positions technology and technologically enabled media platforms in the service of print-based composition—to the point where such a view becomes instantiated as disciplinary prerogative. We find this focus at work, for instance, in Janice M. Lauer's chapter "Rhetoric and Composition" in Bruce McComiskey's 2006 collection, *English Studies: An Introduction to the Discipline(s)*, which serves as an all-purpose guide to the profession. In her chapter, Lauer barely mentions computers and writing, although a brief section, "Computers and Composition," acknowledges that

> since rhetoric and composition's beginning, there has been a focus on writing as a technology. . . . As the computer emerged, and then the Internet, theorists and teachers investigated the relationship between composition and these technologies. The field has also introduced technology into composition classrooms, both in terms of planning, drafting, and revising online, and of participating in chatrooms. (124–25)

In Lauer's words, the "relationship between composition and these technologies" centers on using technology to foster the established writing process. The powerful mode of delivery that Porter references becomes backtracked, a tool to aid invention; put another way, for Lauer, computers exist to serve process pedagogies.

We can catch a sense of such instantiation at the disciplinary and professional level by examining the statements of our professional organizations on the use of new media and the composition of multimedia texts in composition classrooms. For instance, the 2004 "CCCC Position Statement on Teaching, Learning, and Assessing Writing in Digital Environments" acknowledges that "digital writing" can occur in a variety of forms and genres:

> For example, such composing can mean participating in an online discussion through a listserv or bulletin board (Huot and Takayoshi). It can refer to creating compositions in presentation software. It can refer to participating in chat rooms or creating webpages. It can mean creating a digital portfolio with audio and video files as well as scanned print writings. Most recently, it can mean composing on a class weblog or wiki. And more generally, as composers use digital technol-

> ogy to create new genres, we can expect the variety of digital compositions to continue proliferating. (CCCC, "Position")

In light of such "proliferating" texts, the statement's authors claim that "the focus of writing instruction is expanding: the curriculum of composition is widening to include not one but two literacies: a literacy of print and a literacy of the screen. In addition, work in one medium is used to enhance learning in the other." The final sentence is curious in its ambiguity: which medium is used to "enhance learning in the other," or does each enhance the other? The lack of specificity leaves the door open to considering how the new "variety of digital compositions" can enhance the kinds of writing already occurring in composition courses. Under the heading "Assumptions," the authors state unequivocally that "as with all teaching and learning, the foundation for teaching writing digitally must be university, college, department, program, and course learning goals or outcomes" (CCCC, "Position"). The course and institutional goals for digital and new media are prioritized.

What might some of those goals be? The statement is suggestive but not entirely clear. The authors propose that "as we refine current practices and invent new ones for digital literacy, we need to assure that principles of good practice governing these new activities are clearly articulated" (CCCC, "Position"). What might those good practices be? The assumption is that writing instructors will know best. And indeed, the statement asserts that "regardless of the medium in which writers choose to work, all writing is social; accordingly, response to and evaluation of writing are human activities, and in the classroom, their primary purpose is to enhance learning" (CCCC, "Position"). The particularities of the modalities or media used are not addressed; the goal is the enhancement of learning—and we must assume that that learning will follow the aims of compositionists and need not be attendant to either the particular histories of media or the particular rhetorical affordances of the media used. For instance, in the creation of video essays, we are exonerated from knowing anything about the history of film or the rhetorical tropes and figures developed in the history of video production; rather, we can focus on the enhancement of learning in the pursuit of institutional learning outcomes.

To be fair, those outcomes *could* include a historical understanding of the media involved and a required attendance to the rhetorical capabilities of new media. At one point in the statement, the authors acknowledge that "given new genres, assessment may require new criteria: the attributes of a hypertextual essay are likely to vary from those of a print essay; the attributes of a weblog differ from those of a print journal" (CCCC, "Position"). We agree. But note the genres used as examples: the essay and the journal. Again, even when acknowledging that different media forms and genres will "require new criteria"—not just of assessment, but of instruction—the examples offered situate those new media in relation to two of the privileged texts of composition studies, the essay and the journal.

Such reliance on traditional modalities of composing is obvious in disciplinary reviews of our field's assessment practices. Michael R. Neal's "Assessment in the Service of Learning" offers a compelling review of Diane Penrod's *Composition in Convergence: The Impact of New Media on Writing Assessment* (2005), and he shares with Penrod, in her words, a "concern that digital writing and its assessment might lead to a new era of current traditional practices that emphasize form, surface-level errors, and mechanics" (Penrod 43). Why? As Neal argues, the reliance on older formal structures responds directly to feelings of threat:

> When faced with something unfamiliar and uncomfortable, writers and teachers have the tendency to resort to safe places such as current traditionalism, which can be defined and defended in response to a critical examination. That is why defining explicit criteria in terms of new media and other digital writing is so timely at this juncture of teaching composition. (753)

In other words, effective assessment of multimodal compositions cannot rely on older forms and genres; rather, "the rules of the assessment game must change to emphasize rhetorical and linguistic improvisation and innovation as opposed to predetermined, set criteria" (753), and "assessments need to keep pace with new composing technologies rather than trying to affix criteria of print literacies on digital texts and composing processes that are emerging with new media composition" (754). However, we are far from doing this.

The Council of Writing Program Administrators' "Outcomes Statement for First-Year Composition" briefly mentions goals for composing in electronic environments, and their vision for those environments recalls that of the CCCC statement. The outcomes in general focus on rhetorical knowledge, critical thinking, reading and writing processes, and knowledge of conventions—emphasizing throughout the need to acquaint students with multiple genres, but particularly those in which students will work academically and professionally. The final statement, "Composing in Electronic Environments," picks up on this trend:

> As has become clear over the last twenty years, writing in the 21st-century involves the use of digital technologies for several purposes, from drafting to peer reviewing to editing. Therefore, although the *kinds* of composing processes and texts expected from students vary across programs and institutions, there are nonetheless common expectations.
>
> By the end of first-year composition, students should:
>
> - Use electronic environments for drafting, reviewing, revising, editing, and sharing texts
> - Locate, evaluate, organize, and use research material collected from electronic sources, including scholarly library databases; other official databases (e.g., federal government databases); and informal electronic networks and internet sources
> - Understand and exploit the differences in the rhetorical strategies and in the affordances available for both print and electronic composing processes and texts. (Council)

The final bullet point, we contend, moves in the right direction, acknowledging a bit more baldly that different media offer distinct "differences in . . . rhetorical strategies," but the overall situating of this point relegates it to a position below the emphasis in the first two bullet points on using "electronic environments" to facilitate the writing process and to find reliable information, presumably in the creation of composed essays.

These statements make some small gestures toward the particular affordances of new media and multimodal texts. At the same time, however, the statements suggest that these media are essentially tools to be used by compositionists in the pursuit of the development of writing processes for the production of standard essays, discipline-based writing, and professional discourse. Indeed, the concluding comment on the WPA outcomes statement maintains that "faculty in all programs and departments can build on this preparation by helping students learn" both "how to engage in the electronic research and composing processes common in their fields" and "how to disseminate texts in both print and electronic forms in their fields" (Council).

To be sure, we do not believe that these are—in and of themselves—bad goals. In significant ways, they seem completely appropriate for composition courses, which often serve as gateway courses to thinking and writing in other disciplines. Moreover, the techno-inclusionist impulse certainly arises out of a desire to work with the kinds of texts and platforms that excite student interest. However, not all instructors are in a position to learn entirely new fields of discursive, visual, and aural practice. Focusing on what compositionists know and viewing the new media through that knowledge allows us to bring such media in and make sense of it with our students. And writing teachers inevitably have a lot to offer. In *Writing New Media: Theory and Applications for Expanding the Teaching of Composition*, coauthors Anne Frances Wysocki, Johndan Johnson-Eilola, Cynthia L. Selfe, and Geoffrey Sirc grapple with just this point, and Wysocki astutely makes a claim for what compositionists bring to the new media table: "I want to argue that new media needs to be opened to writing. I want to argue that writing about new media needs to be informed by what writing teachers know, precisely because writing teachers focus specifically on texts and how situated people (learn how to) use them to make things happen. Such consideration is mostly lacking from existing writing about new media" (5).

We do not necessarily disagree. But we are even more drawn to a move that Wysocki makes just a few pages later—a nicely historicizing move that opens up, we believe, new possibilities. In talking about looking at the history of media, Wyoscki asserts the following:

> There needs to be more of this sort of critique for new media, which shows us—because of its attentiveness to the particular material ways we use communicational technologies and media—that new technologies do not automatically erase or overthrow or change old practices. If there are openings for change in new media, we can take advantage of them if we are attentive not only to what is new but also, necessarily, if we are attentive to what is old and hanging on (and hanging on, especially, quietly, in places that do not call attention to themselves). This is the kind of work that teachers of writing are prepared to do precisely because of how they see texts as complexly situated practice embedded in the past but opening up possible futures. (Wysocki et al. 8)

Wysocki creates a space in which we—and our students—can begin to experience the play of media across time, media transforming into and out of one another, media informing other media practices, media becoming, in the words of Bolter and Grusin, "remediated." This space already begins to make good on the title of Wysocki et al.'s book: *Writing New Media: Theory and Applications for Expanding the Teaching of Composition.* As the authors make clear, new media prompts an "expanded" notion of composition—one that moves beyond reliance on the essayistic.

The question now is, why? And, moreover, how might such an expansion best be accomplished? Perhaps more basically, what makes this historicizing gesture important?

THE WILL TO HISTORICIZE: THE CHALLENGE OF THE "NEW" MEDIA

Our characterization of techno-inclusionism has thus far focused on a trend in the larger profession and discipline of composition studies to figure the new media as delivery systems for composition's primary practices and texts—namely, process pedagogies and the argumentative, composed essay. Some recent scholars in rhetoric and composition studies, however, have begun to insist that we take a closer look at the *histories* of media production and communicative practices, as well as the value of thinking "interdisciplinarily" about media. In part, this move follows the greater trend in composition studies to historicize our work as a way of understanding what values and ideologies have shaped its current practice. For instance, Adam Banks, Damián Baca, and LuMing Mao trace the development (and elision) of non-Western rhetorics (African American, Mestiza/o, and Chinese, respectively) as a way to gain critical insight into the particular ideological valences of Aristotelian rhetorical practice.[1] Likewise, Susan Miller, in *Trust in Texts,* puts forth her most powerful argument for looking at the West's own forgotten rhetorics. Her book focuses on rhetorics as "metadiscourses that organize persuasion around specific conditions of trust"; they are "infrastructures of trustworthiness we are schooled to recognize" (2). According to Miller, if one explores rhetoric through the lens of different pedagogies of emotion and trust, Athenian *paideia* emerges as only one of a number of such pedagogies. Such a revised historiography, Miller writes, might "encourage writing studies to collect its balkanized emphases on various purposes and genres of writing under a less leaky theoretical umbrella than reapplied oratorical categories offer" (139).

Analyses from scholars Stephen Mailloux and Byron Hawk seem particularly compelling here. In *Disciplinary Identities: Rhetorical Paths of English, Speech, and Composition*, Mailloux succinctly rehearses the various "splits" that have occurred in the formation of contemporary fields and subfields in written and oral communication. He argues that the fragmentation of "rhetoric" into contemporary disciplines such as communication and composition has potentially damaged productive cross-fertilization and delimited our ability to think productively about how a variety of "texts" move in the world:

> Historically, I believe, the separation of written rhetoric (literature and composition) from oral rhetoric (public address and debate) in the academic humanities, this separation of English and speech departments, has been just as debilitating for rhetorical study as the literature–composition split in English departments. One might speculate about how these divisions encouraged the eventual exclusion of oratory from British and American literary canons. More important, these divisions have resulted in a fragmented disciplinary approach to everything having to do with tropes, arguments, and narratives in culture, including most recently the study of local and global communication networks, old and new literacies, and past and present media revolutions. (39)

The gesture at the end of this statement to "global communication networks" is telling, particularly since numerous disciplines—from composition studies to anthropology to informatics—take up and analyze from numerous standpoints the flows of digital communication. Mailloux—and we with him—worry about the discussions that are not taking place due to the fragmentation he describes. What might we learn from one another if we reached beyond our disciplinary enclaves?

More recently, Byron Hawk, picking up on Mailloux's critique, wonders what disciplinary limits are being placed on our consideration of media and communication (broadly conceived) precisely because of our commitment to disciplinarity. In his book *A Counter-History of Composition: Toward Methodologies of Complexity*, Hawk considers in particular the elision of vitalism in composition studies. His overarching critique, however, is that a heavy reliance on Aristotelian logic has hampered our ability to think capaciously about rhetorical possibilities. Hawk argues that Aristotelian conceptualizations of the subject as agent are too simplistic, particularly given the complex ways through which contemporary "subjects" construct and mediate multiple identities:

> The image of a single, central, stable subject gives way to a multiplicity of selves that emerge through complex relations, which goes beyond Aristotle's custom, or dramatis personae. The subject is not simply the political position or identity someone chooses but the relationships established through those identities and the effects of those relationships on bodies. . . . Such an ethics/ethos of networked relations will be a key to future understandings of rhetoric in networked cultures. (189)

Even more pointedly, Hawk implicates composition in failing to address the complexity of such "networked relations." He cites James Berlin in particular as indicative of a compositionist who had progressivist intentions and aims, but whose reliance on older models of rhetorical practice failed to consider the diverse ways in which individuals and groups are called upon to compose, particularly in the public sphere:

> Berlin's social-epistemic rhetoric is still caught in a rationalist epistemology and a dialectical methodology. If rhetoric and composition is to move forward and go beyond the dialectics of social-epistemic rhetoric's model of the communications triangle, it can no longer settle, much less strive for, the production of overly simple heuristics and process models to account for the complexity of invention and writing. (205)

The question remains, however, how composition might "move forward" and "go beyond . . . the production of overly simple heuristics." And more specifically, as we ask students to engage a variety of media platforms for their compositional efforts, and as we, in Mailloux's words, think with our students more acutely about their participation in "local and global communication networks," how do we productively expand students' (and our own) sense of the possible?

Scholars who study new media offer us a potent set of possibilities, particularly as their work in the history of media parallels recent trends in composition studies to historicize our privileging of certain forms of textual production. These scholars recognize the importance of understanding how media emerge at particular moments to do particular work; those histories in turn shape our understanding of those media, as well as our sense of what can and cannot be done with them.[2] For instance, Lev Manovich offers a powerful insight in his masterwork, *The Language of New Media.* Thinking about the histories of new media, Manovich titillates us with some of the reasons why thinking historically about the emergence of media may be important:

> Modern media and art pushed [aesthetic and communication] techniques further, placing new cognitive and physical demands on the viewer. Beginning in the 1920s, new narrative techniques such as film montage forced audiences to bridge quickly the mental gaps between unrelated images. Film cinematography actively guided the viewer to switch from one part of a frame to another. The new representational style of semi-abstractions, which along with photography became the "international style" of modern visual culture, required the viewer to reconstruct represented objects from a bare minimum—a contour, a few patches of color, shadows cast by the objects not represented directly. Finally, in the 1960s, continuing where Futurism and Dada left off, new forms of art such as happenings, performance, and installation turned art explicitly participational—a transformation that, according to some new media theorists, prepared the ground for the interactive computer installations that appeared in the 1980s. (56)

On the one hand, Manovich argues that various media "pave the way" for yet other, newer media to emerge. On the other hand, he also suggests that "new" media essentially teach us different ways to experience what is being mediated. More powerfully yet, the participational media of the 1960s avant-garde prepared us for new kinds of participation in networked cultural spheres that have, Manovich suggests, real consequences for how we interact with, process, and create meaningful data. Put another way, tracing these histories of media shows us how our own interaction with and creation of media shifts, changes, and alters. New media offer us potentially whole new ways to experience the world. These new ways do not invalidate older ways of knowing; people certainly still read books, even in the age of global networked communications. However, the interaction among media is complex, as Bolter and Grusin have noted with their concept of *remediation*, or the retrofitting of older media to look like newer media.

We dwell on this passage because it suggests so powerfully that a pedagogical approach to composing with new media must consider the ways in which developments in media invite composers to become much more cognizant of the forms, modalities, genres, and technologies through which they communicate. The unfolding history of media technologies challenges us to reconceive, again and again, how we communicate with and through media, how media interact with one another, and how we reflexively understand ourselves, individually and collectively, in our interactions across different media platforms. And in this unfolding, we catch a sense of how our notions of rhetorical effectiveness must become increasingly flexible. In *Lingua Fracta: Towards a Rhetoric of New Media*, Collin Gifford Brooke argues for paying close attention to what he calls the "technological specificity of various rhetorics":

> I believe that, as teachers and students of writing, scholars in composition and rhetoric are indeed uniquely positioned to contribute to discussions and debates about new media. Such contributions, however, depend on our ability to rethink some of our own cherished and unexamined assumptions about writing; new media will transform our understanding of rhetoric as thoroughly as our training and expertise in rhetoric can effect a similar impact in discussions of new media. But this presumes that we recognize the various contributions that information technologies make to rhetorical situations. Rhetoric and composition has emerged over the past century within the disciplinary context of the study of literature, and that context has predisposed us to print literacies and textualities. That predisposition is strong enough that we tend to neglect the technological specificity of various rhetorics,

> and this, in turn, has kept us from bridging the gap. . . . This gap, between the local particularity of the individual text and the global generality of media structures, is a space that we already occupy as writers and writing scholars; bringing what we know to bear on new media is the next logical step in the growth of our discipline. (5)

To "recognize the various contributions that information technologies make to rhetorical situations" is certainly one of the central goals of Brooke's book—and ours. We need not rehearse entire histories of media to recognize that predispositions to particular forms of communication might prompt us to "neglect the technological specificity of various rhetorics." But a *historicized* sensitivity can show us how we might, when thinking and composing with new media technologies, "rethink some of our own cherished and unexamined assumptions about writing." Writing has been valued differently across time, been made to do different work. What work might the new forms of composing do—now and in the future?

Before we proceed further, we must take stock of one absence in Brooke's formulation. Brooke notes what he calls a "gap"—specifically "between the local particularity of the individual text and the global generality of media structures," and he claims that this "is a space that we already occupy as writers and writing scholars." What seems missing in Brooke's formulation is a consideration of what may lie within that gap (besides us as writing teachers). Specifically, Brooke does not offer us an accounting of public spheres, which mediate individual (and collective) authors and texts with(in) larger structures of meaning-making. Theorizing the public sphere(s) allows us to reposition ourselves as writing instructors. Moreover, considering the fact that many public spheres are now themselves complexly mediated through new media platforms, a failure to think through the rhetorical demands and possibilities of such public spheres seems damaging to any pedagogy we might mount. So we turn our attention now in the final section of this chapter to a consideration of public spheres and their relation to new media in an attempt to think critically about our role—and our discipline's role—in composing with new media.

INTO THE PUBLIC SPHERE

Why does our understanding of media and composing with new media platforms need an expansiveness beyond the kinds of compositional practices most cherished and promulgated in composition courses? Specifically, the current practices of techno-inclusionism are insufficient pedagogically for our students' needs; the changing nature of communication in many public spheres demands a greater attention to modalities of communication that exceed the essayistic. That is, in addition to being able to write professionally and compose credible letters to the editor or op-ed pieces, students need to be able to participate actively and with good ethos in vibrant public spheres that are often encountered and structured through multimedia and multimodal forms of communication. Indeed, having moved from sounding the warning notes of imminent technological threat and the need to "pay attention," Cynthia Selfe acknowledges in her chapter of *Writing New Media*, "Students Who Teach Us: A Case Study of a New Media Text Designer," the increasingly prominent connection between multimedia composing and participation in public spheres. She writes: "If educators hope to prepare citizens who can 'participate fully' in new forms of 'public, community, and economic life'—in other words—we must teach them to design communications using modes of representation much broader than language alone" (55).

Such thoughts parallel Selfe and Hawisher's findings in *Literate Lives in the Information Age*, a detailed and multiyear study of the proliferation of different kinds of technologically enabled literacies. Their study, undertaken through extensive surveying and analysis of numerous case studies, highlights some characteristics of contemporary writing and literacy practices, particularly as these have been inflected and altered by the use of new communications technologies and multimedia formats. They conclude the following:

> Faculty members, school administrators, educational policy-makers, and parents need to recognize the importance of the digital literacies that young people are developing, as well as the increasingly complex global contexts within which these self-sponsored literacies function. We need to expand our national

> understanding of literacy beyond the narrow bounds of print and beyond the alphabetic. (232)

A clearer call we cannot imagine. Our curricula must change in order to meet the demands of a significantly changing global public sphere, connected through multiple technological and digital interfaces. In contrast to professional and organizational statements, Selfe and Hawisher assert the need to prioritize exploration of the kinds of literacy practices that students are developing, as opposed to front-loading our composition curricula with our particular course and institutional aims, goals, and desires. If students are not just to succeed in particular job or career contexts, but also to participate in different electronically enabled public spheres, they need a significantly more complex understanding and appreciation of how multimedia and multimodal communication function.

Jody Shipka's *Toward a Composition Made Whole* perhaps comes closest to arguing for a radical expansion of what we consider composing and what we teach as "writing." In her conclusion, she maintains that " we must find ways to underscore for students what has always been the case—that communicative practices are multimodal and that people are rarely, if ever, just writing or making meaning with words on a page" (138). Shipka seems to understand our fixation with textual production as the primary problem, and she notes how she has "warned against research and pedagogical frameworks that overlook, or worse yet, render invisible the complex and highly distributed processes associated with the production of texts, lives, and people, thereby obscuring the fundamentally multimodal aspects of all communicative practice" (144). Situating her position squarely in the conversation that reached a critical level with the publication of Selfe's 2009 essay on sound ("Movement"), Shipka calls for an opening up of compositional practice:

> To be clear, in suggesting that students be provided the option to accomplish academic work via the employment of representational forms, genres, or modes that are not typically associated with that work, my intent is not to demonize or downplay the value or import of linear, thesis-driven, double-

> spaced alphabetic texts, texts that largely resemble, well, this very book, in fact. With a mind toward a concern raised by Doug Hesse (2010) in his response to Cynthia Selfe's 2009 "The Movement of Air, the Breath of Meaning," students with whom I work always have the option to explore "new ways of making meaning" that "include writing extended connective prose" (Hesse 2010, 605). What is most important is, first, that students come away from the experience of the course more mindful of the various ways in which individuals work with, as well as against, the mediational means they employ. Of equal importance is that students can articulate for others the purposes and potentials of their work. My hope is that students will continue to choose wisely, critically, and purposefully long after they leave the course—that they will continue to consider the relationships, structures, and representational systems that are most fitting or appropriate given the purposes, potentials, and contexts of the work they mean (and in other cases, *need*) to do. I also think it is important that we challenge students and that we challenge ourselves—whether this involves taking risks and trying something new or considering the various ways in which meaning (both within and beyond the academy) might be accomplished. (145)

We quote from Shipka at length here because her work articulates clearly the importance of considering multimodal compositional practices as enabling more robust participation in complex public spheres. We couldn't agree more. But what do we mean by complex public spheres?

The work of Michael Warner shapes, in part, our understanding of the needs of these spheres. Warner, writing with Habermas in mind, attempts to theorize a rhetorically active public sphere, one that considers issues and debates them vigorously—if not always as "rationally" as Habermas intends. Indeed, Warner argues:

> Public discourse . . . is poetic. By this I mean not just that it is self-organizing, a kind of entity created by its own discourse,

> or even that this space of circulation is taken to be a social entity, but that in order for this to happen all discourse or performance addressed to a public must characterize the world in which it attempts to circulate and it must attempt to realize that world through address. (113–14)

Put another way, for Warner, a "public is poetic world making" (114), grounded in discourses that attempt to frame and figure what the interlocutors believe to be true about the world. What is deemed "true," though, is debatable and not, à la Habermas, grounded in rational debate; indeed, for Warner, the circulation of discourse in the public sphere does not proceed through a series of rational assertions that are discussed, debated, countered, and reconceived. Rather, Warner envisions the assertion of lifeworlds that may or may not be rational but that certainly exert authority and norm behavior. One might attempt to argue rationally with such lifeworlds, but Warner argues that more effective public "debate" occurs through iteration and citation: "Success in [the public sphere] is not a matter of having better arguments or more complex positions. It is a matter of uptake, citation, and recharacterization. It takes place not in closely argued essays but in an informal, intertextual, and multigeneric field" (144–45). In this public sphere, "intellectuals," as Warner identifies them and by which he means scholars, academics, and the intelligentsia, are not necessarily better equipped to put forth arguments and be persuasive. Such rationalist approaches are not useless; rather, Warner believes that powerful norms, affects, and belief systems—all of which might exceed the rational and the rationally arguable—are picked up, circulated, reiterated, and recharacterized through multiple modalities and genres to reinforce those norms, affects, and belief systems. Being attentive to the extra-rational and the circulation of normative beliefs, then, is key to entering successfully into the "conversation" of the public sphere.

As an example of how such a view of the public sphere plays out, we might consider Warner's distinction between publics and counterpublics—those whose lifeworlds dominate the public discourse versus those whose lifeworlds do not, such as heteronormativity versus queer subcultures. As Wallace and Alexander explain in "Queer Rhetorical Agency,"

> agency in the public sphere relies on being able to articulate and *maintain* the dominant lifeworld. Such explains the high level of activity around gay marriage issues, for instance; opponents of gay marriage are attempting to preserve a particular view of what marriage is and how it should be defined, while proponents argue that gay relationships are often just like marital ones, so they should be labeled and understood as such—not just rhetorically, but in material reality. (817)

Queer selves often only gain legitimacy in the larger public sphere to the extent that they mimic, as pale imitations, the heteronormal—we want to get married (check normal), we want to serve in the military (check normal), we want to join the status quo, helping to maintain it in the process. Hence the plethora of commercials and ads in support of gay marriage that show how gay couples are much like straight couples—they go on dinner dates, they do laundry, they raise children, they enjoy and suffer the ups and downs of a life course. The iterative and metaphorical nature of arguments for gay marriage—"See, it is like straight marriage"—is certainly part of what leads Warner to argue that the public sphere proceeds more poetically than rationally. The hope here is that such citationality will not merely repeat but also alter in its repetition the construction of marriage in the public sphere; the hope is that marriage will become a less exclusive, more expansive marker of social standing. However, the entrance into the debate is enabled almost exclusively by the initial gesture—the rhetorical move—of iterability.

Warner's view of the public sphere is hardly totalizing. The public sphere is most likely diverse and complex enough that it proceeds both rationally and poetically at the same time—and often with both in tandem. We do not mean here to set up a binary. Indeed, ancient rhetorical traditions may have privileged logos, but they also were well aware of the necessity of appeals to the "irrational" or nonlogical, and the overlapping of rhetorical and poetic figures speaks powerfully to the "aestheticization" of rhetorical practice. Our intent, however, is not to rehearse this particular history but to note that if we are invested in preparing students for active participation in different public spheres, we may very well need to equip them with a range of strategies—strategies that address both the Habermasian rationalist and the intercommunicative public sphere and strategies that engage the more poetical, citational, and iterative public sphere that Warner identifies.

How might we best prepare our students—and prepare ourselves—for rich participation in this complex public sphere? And how might we teach (with) the new media to equip students for both a rationalist and a poetical understanding of the public sphere? Unsurprisingly, Selfe and Hawisher point the way at times. Near the end of *Literate Lives in the Information Age*, we find a common refrain in the scholarship about the need to understand literacy contextually: "Literacy exists within a complex cultural ecology of social, historical, and economic effects. Within this cultural ecology, literacies have life spans" (212). And more recently, in a 2011 *CCC* article, "Ecological, Pedagogical, Public Rhetoric," Nathaniel A. Rivers and Ryan P. Weber argue along nearly parallel lines that "public rhetoric scholarship and pedagogy could benefit from an expanded scope that views rhetorical action as emergent and enacted through a complex ecology of texts, writers, readers, institutions, objects, and history" (188). Similarly, David Sheridan, Jim Ridolfo, and Anthony J. Michel have well articulated the deeply intertwined relationship between fostering multimodal composing and promoting participation in public spheres. In "The Available Means of Persuasion: Mapping a Theory and Pedagogy of Multimodal Public Rhetoric," they argue for a "rhetorical education" that

> has an important role to play in fostering . . . public sphere participation. In order to take advantage of the special cultural work that multimodal rhetoric can perform, students, as public rhetors, need to be given opportunities to produce and repurpose multimodal compositions that counter existing hegemonic vocabularies. Doing this means confronting such questions as, What is the nature of the exigency that calls

> for an intervention? What rhetorical practices contribute to this problem? How is it reinforced and reproduced by the circulation of images, metaphors, stories, and representations? What kinds of counter-practices can effectively intervene? What new images, metaphors, stories, and representations need to be placed into circulation if consciousness is to be altered? How can hegemonic naming and framing practices be destabilized? How can new naming and framing practices be introduced? (829)

We thrill to precisely the same questions, believing that students should be afforded the opportunity to engage complex multimodal compositional spaces so they might map out their own counter-publics, questioning, querying, interrogating, and intervening.

At the same time, we note the tendency to favor and privilege the "new." Even in Sheridan, Ridolfo, and Michel's superb call for a multimodal rhetorical education, we hear echoes of "new images, metaphors, stories, and representations," as well as "new naming and framing practices." Of course we will have new images and frames. But what about the *historical* contexts of such practice? They garner a mention, but what would a more historical understanding of technologized communication look like? In *A Better Pencil: Readers, Writers, and the Digital Revolution*, Dennis Baron justly deflates the promise of the "digital revolution":

> It turns out that the digital revolution is playing out as all communication revolutions do. Computers don't live up to the grandiose promises of their biggest fans. Nor do they sabotage our words, as critics loudly warn that they will. Instead, as we learned to do with earlier writing technologies, once we adopt the computer, we adapt it to our needs, and along the way we find new and unexpected ways of changing what we do with words, and how we do it. (xvii)

While we agree, we note the continued "forward look" in Baron's summation: we will, of seeming necessity, find "new and unexpected ways" of doing things with words. True enough. But we also want to call attention to the many ways in which our fascination with the "new and unexpected" overlooks a variety of powerful rhetorical affordances from past revolutions, from older ways of doing things with (and to) words, images, sound. We also need to be attentive to the historical *now*, the many ways in which current sociocultural, political, and economic forces shape our understanding of media, and ourselves in and through media.

Perhaps what our field most needs at this moment—at the moment that it embraces a variety of media platforms for compositional work—is a simultaneously *historicized and poeticized* understanding of new media. Simply put, students will have a greater ability to compose with new media if they understand the particular rhetorical capabilities of the media platforms they use. One powerful way to help students develop a sense of those capabilities, we argue, is through a consideration of different uses of media across time and in different contexts. Such an expanded history of new media shows us unequivocally that it has often been used not just in rationalist, essayistic ways, but also in more poetical ways that emphasize iteration, citation, parody, and pastiche—the potentially rich rhetorical tools that Warner identifies as essential to robust participation in the contemporary public sphere. In the following chapters, we explore an array of rhetorical affordances that may enliven and expand our notions about the capacities of the new media to engage and debate issues in complex public spheres.

2

Direct to Video: Rewriting the Literacy Narrative

As early as 2003, Melissa Meeks and Alex Ilyasova could survey the "new and exciting" use of digital video in composition classrooms, noting the sorts of multiliteracy that Stuart Selber advocates and engaging students in prosumer activities. Digital video production has, according to Meeks and Ilsayova's interviewees, a number of benefits, including its power "to engage many literacies at once"; its stimulation of collaboration and participation; and its involvement of "students in a rich composition process" (Meeks and Ilsayova). More recently—and these are only a few examples—Jeannie Parker Beard describes making use of cell phone video and YouTube, among other technologies, for a first-year writing class's final project, "proposal documentaries" (2010); Claire Lutkewitte explores a variety of Web 2.0 technologies (including YouTube) with her first-year writing classes; and in an installation for the 2010 Computers and Writing Conference, Bill Wolff showcased a year's worth of student videos created in an upper-division writing course, videos that challenged viewers "to rethink traditional concepts that so often seem fixed in meaning and performance: text, research, writing, composition, among others." Indeed, this compositing modality is deeply rhetorical, as Gunther Kress notes in his discussion of multimodal design, which works "to present, to realize, at times to (re-)contextualize social positions and relations, as well as knowledge in specific arrangements for a specific audience. At all points, design realizes and projects social organization and is affected by social and technological change" (*Multimodality* 139).

We agree. However, as we consider composition courses that experiment with and include video production as part of their work, we note that some anxieties about media production often manifest as a privileging of traditional textual modalities of meaning-making, even as instructors (seemingly) embrace newer forms of delivery. In this chapter, we offer our thoughts on how composition's embrace of digital video often invites students to participate in the production of multimedia texts but, at the same time, often separates those texts from a robust consideration of the rhetorical affordances of video. In the process, we argue, students'—and our own—understanding of multimedia, multimodality, and digital composition is impoverished and emptied of much critical and rhetorical possibility. As Meeks and Ilyasova's interviewees point out,

> *Knowing rhetoric may not license us to create and critique anything and everything.* . . . [Film] studies and film production have bodies of knowledge and sets of inquiry tools all their own. Though rhetoric can certainly facilitate the effective use of digital video in the classroom, we must begin to be more intentional in our borrowing from the professional programs and academic disciplines that have been using these media longer, with more sophistication. (Meeks and Ilsayova; authors' emphasis)

We argue that preparing our students for literate participation in complex, multicultural public spheres may very well mean equipping them with this more robust vocabulary of textual, visual, and multimodal meaning-making—a vocabulary that should also include the nonrational, the alternative, the knowledges of the body, and the avant-garde as part of its critical lexicon.

To be clear: our goal here is not to argue against including digital video production in the writing classroom; we believe—and assume a belief in our readership—that engaging multimodality is a pressingly necessary task for a wide variety of composition and writing studies courses. Video composing has become a key modality of meaning-making among younger generations of college students, so developing a critically literate approach to such textual production seems crucial. At the same time, we question our field's approach to and use of such texts in the composition classroom.

How critical and complex is that approach? To what extent do we ignore a rich history of multimedia in order to colonize the production of such texts with our compositional aims, biases, and predispositions?

VIDEO AS REPLICATION OF THE ESSAY

Innovative programs in writing studies foster growing interest in the creation of video "texts," providing computer access, software, and instruction. At the University of Texas, for instance, *TheJUMP*—the *Journal for Undergraduate Multimedia Projects* (jump .dwrl.utexas.edu)—is an online resource and repository for students' multimedia work. According to *TheJUMP*'s website, the journal is dedicated to:

> providing an outlet for the excellent and exceedingly rhetorical digital/multimedia projects occurring in undergraduate courses around the globe, and to providing a pedagogical resource for teachers working with (or wanting to work with) "new media." The journal is designed to be not only a repository for quality multimedia scholarship—bringing together some of the most rhetorically creative and rhetorically impactful works produced/composed by our undergraduates—but also, unlike its digital brethren (i.e., megarepositories like YouTube), it seeks to also offer a critical perspective.
>
> As such, the projects we publish include assignment descriptions from the courses in which they originated, reflections by the instructors involved, and design rationales or process/product reflections by the author(s)/composer(s) themselves. In these reflection pieces, the creators attempt to critically consider their design/production choices and/or the intent of their projects in light of their rhetorical message, their "composing" process, and the technologies involved.

As a resource for faculty interested in learning how to design multimedia assignments, *TheJUMP* offers some of the most tried-and-true aspects of composition pedagogy, including the "meta-reflection" or "writer's memo" in which students discuss the particular rhetorical choices they made in the composition of their multimedia texts. As a showcase for student work, *TheJUMP* presents viewers with an array of creative projects, experimenting with audio, visual, and textual production in often highly effective—and affecting—ways.

As just one example among many, the short film "Closer," by Kyle Kim (jump.dwrl.utexas.edu/old/content/kk), presents a young man playing chess and eating a bagel alone in a cafe. He gets up to get a drink and a young woman sits down at the chess table. Instead of joining her, he moves to another table to await her eventual departure. The video has no dialogue or narration, save for a movingly performed song in French ("Une Nouvelle Histoire" by DoKashiteru). An accompanying text reveals that the video "was part of the course's final project assignment of the intensive 'January Term' at Whitworth University, and it asked students to find a way to merge audio, video, and text into a coherent narrative." Essentially a music video, "Closer" quite effectively generates a pathos-laden argument about what it means to interact and become "close" to others. However, in Kim's reflection on his work, he notes the ways in which his thinking about the composing process for his video benefited from thinking critically about past experiences in purely *textual* production:

> Most journalism students are taught to craft stories within a more traditional medium of print (with words or through photojournalism). Experiencing a process that was more artistic than creating journalistic prose made me compare and contrast different methods and approaches with the kind of storytelling I am used to doing. In journalism, reporters are ethically limited as to how a story is carried out (research, interviewing and writing style, to name a few of those limitations). With my short fictional film, I found myself given more leeway in tone, style and implementation; bias, fairness and objectivity were irrelevant issues in this case. It is even arguable that in fiction bias is an intricate component of the story (e.g., point of view/narration, character portrayal,

> theme and representation). Most common biases in journalism come from taking quotes and facts out of context. In terms of video, post production is the area where bias can easily creep in. How scenes are edited and spliced together naturally creates a certain point-of-view. However, this project required intentionally emphasizing a point of view, and it was interesting to produce with journalistic training that has taught me to combat such biases.

Such reflection reveals a laudable critical engagement with the particular rhetorical capabilities—and potential pitfalls—of different genres, contrasting the need for critical awareness of bias in journalism with the greater "leeway" given bias in fiction. Moreover, Kim notes that the *material* processes of creating the video—"How scenes are edited and spliced together"—become an integral part of creating a point of view.

Certainly, Kim seems to have learned about different modalities of media production, and his instructor seemed quite pleased with the results, writing that the video "is a fine example of both a desired product from the course and a desired process from the course, contributing to Kyle's sophistication as a reader of multimedia compositions and to his expertise as a sophisticated, ethical maker of such compositions for the world of online journalism." The instructor's aims in the original assignment were admittedly open-ended:

> Earlier projects consider the joining of images and words, the ways that visual design conveys meaning, the capture and editing of audio, and basic cinematography. The challenge of the final assignment is to pull together as many of those strands as possible into one coherent package that will include, at least, images, sounds, and text.

This assignment leaves a great deal of wiggle room in which the student can experiment. Curiously, however, the instructor's potential textual biases find perhaps some expression in one paragraph worth quoting in full:

> Viewed in terms of the basic project requirements, "Closer" is undeniably excellent but—skewed as it is toward cinematography—might be called weak in its integration of text. That shortcoming vis-à-vis the basic instructions highlights the need for flexible requirements in multimedia assignments, which can move in unpredictable directions during production. In this case, Kyle talked with me about the project as he developed it, and I could see that what he had in mind would be both a suitable capstone for the course and a valuable exercise for Kyle as a journalist. Kyle thought (correctly) of his tasteful credits and titling as a way to include text, and the harmony of those texts with the other visual and aural elements here speaks to his grasp of visual design. He also included a shot of a novel by Ken Kesey (best known for *One Flew over the Cuckoo's Nest*), and that strikes me as another thematically appropriate use of text. While the "perfect" response to this assignment would likely do more with text, the assignment's parameters allow me to reward the learning, creativity, and labor behind a piece like "Closer," and the rubric makes it possible for a project to be weak in one area and still score quite well overall.

Despite the instructor's call for more "text," he notes that such assignments require greater flexibility in assessment; the nature of experimenting and producing such work, according to the instructor, might move one in "unpredictable directions"—that is, away from the standard kinds of textual production that this instructor is more comfortable assessing.

The editors of *TheJUMP* note in the "About" section on their site that such unpredictability forms a key part of work in multimedia production. One aim of *TheJUMP*, they write, is to create dialogue and discussion about what they call the "murkiness" of this work: "The pedagogical focus of this e-journal is critical to its success as we not only want to see really great projects and the assignments/prompts (and courses) that gave them shape, we also want to consider and work through the nuances of critique, assessment, impact, and so on (often the more murky areas associated with digital multimedia productions)." For us, that murkiness manifested itself in the assessment of "Closer." Some of that murkiness stems from the sheer plethora of video possibilities; for instance, other videos in the same "issue" of *TheJUMP* in which Kim's video appears vary markedly in genre and tone. Whereas Kim offers us a music video, for example, Sarah Gould offers "A Closer Look into Physical Disabilities: An Oral History Video," which functions more like a documentary video project. Further exploration of *TheJUMP* reveals even greater generic divergence—which is only to be expected as students play with the possibilities of video production.

The murkiness that Kim's instructor and the editors of *TheJUMP* point to is perhaps more their problem than the students'. After all, Kim himself notes that working with fiction and working with video are *different*. Certainly, what is at least partly at play here is the move from nonfiction to fiction as genre. As Kim points out in his insightful reflection, however, what is also at play are different modalities of editing, of actual *production*, that affect how one composes. By comparison, his instructor's commentary about the "unpredictable directions" of such production seems, well, old-fashioned, even conservative. To his credit, the instructor allows his writing students to experiment with multimedia; in fact, it's a requirement for the course. However, his *thinking* about such multimedia seems to rely more on privileging some media rather than just distinguishing between them. This kind of thinking favors the textual over other forms of communication; note, for instance, how he calls Kim's work "skewed" toward "cinematography." In other words, this work—as "undeniably excellent" as it is—isn't really writing, and that's a problem—for the instructor. Kim understands that distinction, as articulated in his reflection, and the instructor is rolling with the changes, as it were. But we hear some potential anxiety in his recognition of the "unpredictable directions" this work might take.

Such anxiety is certainly understandable. More traditional modalities of literacy and meaning-making seem to be shunted aside for the newfangled, glitzy multimedia. We are not concerned here, however, with those arguments, and we assume that, given sufficient time, literacy practices just change. And we cannot stop them from changing, so we attempt to offer students a diversity of literate practices and ways of knowing. We *are* concerned with how we as literacy instructors and writing teachers adapt to such changes, incorporating the new literacy practices into our curricula, our pedagogies, and our understanding of what it means to be literate—our *ideologies of literacy*. How do we react to and understand the "unpredictable directions" of incorporating new media into our writing classrooms?

One response to the unpredictability of video is to use it in service of more traditional, writerly composition. We have all probably seen a popular assignment—the video literacy narrative—that has produced any number of visual vignettes through which students learn new technological skills and discover something important about the development of their own and other people's literacy practices. One syllabus presents the assignment thusly:

> For this assignment, you have the opportunity to argue for a particular understanding of literacy by telling a literacy story and then justifying that story's academic rigor. Literacy narratives, as the Digital Archives of Literacy Narratives explains, are stories about reading (books, cereal boxes, music, websites, magazines, signs) and composing (letters, Facebook pages, songs, maps, blogs, papers) in any form or context. Literacy narratives often include poignant memories that involve a personal experience with literacy. Digital literacy narratives are the same kinds of stories told through the use of digital media (iMovie, MovieMaker, Sophie). (Gogan)

Several years ago a number of these video narratives were on display at the 2008 Watson Conference in Louisville, to great acclaim. Increasingly, time and space are set aside at our national conferences and on our campuses to present, view, and laud work produced in response to such assignments—work that reflects on issues of textuality and communication in rich digital formats. For example, UC Irvine recently sponsored, as part of its yearly writing awards, a new award for "Best Multimedia Text." The winning entry, "On

Bad Language," was a video essay about how foul words have rhetorical uses, even if such uses are sharply context-dependent. The author reflects critically on her own literacy practices and analyzes a series of comedy routines and news reports in which foul language serves a strong rhetorical purpose. Clips from routines and reports appear as supporting examples. Note how we describe "On Bad Language" as a "video essay" with clips as "supporting examples." The video essay is perfectly fine in its own right. However, many such texts are overwhelmingly linear in structure, with stated theses and expository narratives, and occasionally with obligatory references to experts. In short, they transport all of the elements of more traditional print texts into another medium, another modality of delivery.

We pause to wonder what else can be done—rhetorically—with this assignment, with this call for students to think about literacy through the delivery of new media. Do such assignments simply replicate essayistic forms in new media? Look back at the assignment, which baldly states that digital literacy narratives are the "same kinds of stories told through the use of digital media." We argue that such a formulation elides a rich consideration of the canon of delivery and its potential impact both on how we understand literate and communicative action and how we represent such action. Put bluntly, we believe that a good deal of contemporary composition practice uses new media and new media tools to replicate and reproduce some of its own cherished forms and genres. We focus here in particular on the video literacy narrative because of its prominence in so many composition programs, which see in it a way to bridge textual and multimedia literacies.

To catch a broad sense of what kinds of videos are produced in comp courses, we sampled a number of videos, working with our graduate students to collect readily accessible video literacy narratives on YouTube, where they are often posted. Two caveats about this survey are vital to consider here. First, while our sample is relatively small, only a hundred of the thousands that exist, we believe it indicates—and informs—the kinds of work produced in response to the video literacy assignment. Second, and more important, we recognize that not all such assignments are the same; some require greater depth of thought and use of available technologies than others. We did not have access to the original assignments that produced these videos, and we do not know the "grades" that students received on their videos. Therefore, we cannot measure the quality of the videos in relation to either the specific assignment or the rubrics used in the course contexts in which they were produced. So our assessment of the students' work might seem unfair. But our goal is not to assess or critique the individual student work, but rather to catch a glimpse of what kinds of videos students produce in writing courses. The sample is necessarily biased in that not all courses require their students to post the videos to YouTube or other publicly accessible forums. We want to know, however roughly, what kinds of qualities characterize students' work with video literacy production—at least as revealed through a sampling of such videos posted on YouTube. And while we cannot generalize from our sample to all such work with video in the field of composition studies, we are nonetheless struck by some consistent strains and predilections in our sample.

As one example, Eric Wooten's "Literacy Narrative" (www.youtube.com/watch?v=6QOgUsHEQUk), posted on YouTube and tagged as "My Literacy Narrative project for my Comp 210 class," shows a student (presumably Wooten) in a series of still images accompanied by lively music from a Beethoven symphony. Textual tags appear throughout the video, tracking the student's reactions to what he sees. He looks for something to read, checking out books on the shelves in the library and deeming most of what he finds "boring." Suddenly, he sees in the library a set of newspapers and magazines about football, and voilà: "Reading is fun," he declares, looking into the camera and giving us two thumbs up. Admittedly, as he puts it, reading other books does not interest him as much as reading about football, but still: "Reading [with some caveats and qualifications] is fun."

Two further aspects of this video are worth comment. First, note how the video proceeds. It presents us a problem, shows us the student trying to solve that problem, and then presents a solution. The linearity of the video isn't itself remarkable, particularly given that videos unfold in linear time. This video, however, really gains nothing by being a video. Its argument could just as easily have been rendered as a three-paragraph theme. In fact, one of the surprising elements of so many of these videos is that they are essentially short themes (often of the five-paragraph variety) delivered via video, with little attention to the rhetorical affordances of video production. Certainly, Wooten uses lively and "triumphal sounding" music; certainly, he mimes the emotions of boredom and tedium and eventual joy that lend pathos to his argument. The overall effect, however, does not compel the viewer. Second, note Wooten's own seeming resistance to the project. His enthusiastic thumbs-up at the end of the video reads more like sarcasm than anything else—sarcasm perhaps about the rather bland thesis that "Reading is fun" but also about the lackluster quality of the video. One is tempted to read Wooten's hyperbole as critique here: he must know at some level that this use of video to vaunt the values of a traditional literacy modality somehow misses the point of using video in the first place. In that context, the response commentary is fascinating; one viewer writes, "Awesome, it really is a literacy narrative." We can imagine a course context in which students post such assignments to YouTube and are instructed to peer review one another's work by commenting on them. The irony of this particular comment is that the video really is just a "literacy narrative"—not a "literacy video." Perhaps we underappreciate students' ability to understand the limitations of such assignments; the poster, after all, might very well be commenting sarcastically that Wooten has fulfilled the assignment but not necessarily made a compelling video—at least not yet. As such, the context of its production, the composition class, might work with video, but it isn't yet mining the rhetorical capabilities of video work.

Admittedly, Wooten's literacy narrative is one of the simpler videos we sampled, in terms of both concepts presented and media used. Other videos from our sample were more adventuresome in their use of media effects, but a set of common practices emerged. Other videos pick up on Wooten's strategy of linking pictures but do so in a more engaging and compelling way. These videos often seem like music videos, with a driving sound track (generally one song) accompanied by fast-moving pictures and textual snippets to guide viewers in terms of what they see and how they might interpret what they see. Alas, such videos, particularly those that seem to be produced in comp courses, are not always as compelling as Kyle Kim's "Closer," which was produced in an upper-division journalism course (not a general education course). For instance, beautifulataxia2's "Digital Literacy Narrative" (www.youtube.com/watch?v=vofvV1C3BUE) presents several different songs melded with images to trace one student's literate development. The second half of the video is dominated by the scrolling text of white print on a black background of a story the student wrote—almost as though the point of the video is to highlight, if not indeed feature, a purely textual modality.

Another common practice is the documentary-style, interview-driven narrative in which the beat-heavy sound track gives way to talking heads that comment on the student composer's development of literacy practices. Parents, siblings, and teachers comment on how the student became literate. The examples and comments seem positioned to reveal a clear trajectory from illiteracy to full literacy. A comparable kind of video is that dominated by just one talking head, in which the student simply narrates into the camera her thoughts about the topic at hand—in this case, how she became literate. In one example of this style of video, Lindsay F's "Literacy Narrative Project" (www.youtube.com/watch?v=HQXltgmXcFw), the composer announces at the outset that her video is not a "written" project but a video because she wants to emphasize the value of multimodality. She describes a course context in which she was asked to read various authors who have written about multimodality and multimedia literacies. Launching into a series of examples, she complains about "summer reading lists" and writing book reviews (cue dramatic music) but then lauds the possibilities of using Facebook and making videos. In one spectacular moment, she pronounces that Deborah Brandt wrote a lot of "gibberish" about literacy sponsorship, and that Brandt's real point is that we should just "go with the flow." That is, we should embrace the different kinds of literacy practices that surround us—such as making videos. The startling thing about this video is that it praises the creation of videos as a powerful literacy practice while the entire video essentially consists of the student rambling through a prepared outline of talking points; the video dimension exists only to deliver this talking head. Granted, that talking head is praising multimodal literacy engagement. In one telling moment, however, the student actually stumbles over her words while talking about the value of engaging different kinds of media to support different kinds of learning, saying, "This is why I am writing . . . not really writing . . . I am telling you in a video." The slippage is understandable; the student is essentially talking her way through an essay about multimodality, as opposed to delivering or documenting her points multimodally.

The talking-head video certainly seems an underdeveloped use of video, and many (fortunately) are not like this. However, a dominant practice in the videos we sampled is not far removed: the voice-over narration with accompanying visuals. While Wooten's "Literacy Narrative" is not heavily narrated (we only "hear" Wooten's thoughts through textual snippets commenting on what's boring, what's not), many literacy narrative videos use voice-over narration as a leading feature. In such cases, the student usually reads a prepared text that is then illustrated with pictures and, in more advanced cases, moving images. For example, Richard Rodriguez composed a video on literacy and video games (www.youtube.com/watch?v=T1z2JkEZJss) that consists of a voice-over narration accompanied by pictures of different video games that he discusses. The overall thesis of the video is that these video games, which Rodriguez played while growing up, assisted him in learning how to read; specifically, beyond the text included in the games, the games' appeal as fantasy adventures led him to read fantasy and science fiction stories and novels. The primary evidence is a series of different examples that cumulatively accrue to support his thesis. Another such text, apesmen09's "Multimodal Literacy Project" (www.youtube.com/watch?v=raJ_eznaDLo), shows us how one young woman actually learned to write—physically. The narrative proceeds as a series of examples of different kinds of actual writing, from printing, to writing in cursive, to composing résumés. Eventually, the composer waxes fondly about her composition class and her development of rhetorical skills, such as thinking about pathos and considering audience. Visuals serve again to illustrate major examples. In cases like these, which constitute a hefty chunk of the videos we sampled, the textual narration dominates, with videos only "accompanying." An essayistic kind of literacy is privileged here, with students (largely) composing narratives first and then *illustrating* them with visual tools. Little attempt to think through the particular affordances of visual narration seems evident. Even more telling, though, is the sense that if the textual narrations were presented alone, they would probably not be assessed very highly; they are generally fairly weak statements about literacy followed by strings of examples. We suspect, however, that the addition of the visual dimension is often evaluated as appropriately enhancing.

Our sampling is hardly exhaustive, but it is representative. Again, we do not know their original course contexts or what grades the videos received. Still, most of the videos are marked by their posters as class related, and those classes are dominantly writing or composition classes. How *do* teachers evaluate such videos? Do such videos represent what is considered "passing"? Regardless, the consistency of qualities and characteristics among them is startling. These videos function as illustrated essays—again pointing to the privileging of essayistic literacies, either among the instructors who assign video production or among the students who respond to prompts calling for video production.

We call out these particular videos not because they are especially bad but because they are especially indicative of the kind and quality of video produced for first-year and beyond composition courses that promote the production of video literacy narratives. We're not claiming that the development of such skills and such narratives and arguments is "bad." In fact, some of the texts are pretty good, at least in terms of mimicking rational textual argument in new media forms. But take Wooten's video literacy narrative as emblematic. On the one hand, the video in many ways seems perfectly fine, piecing together images, music, and text to praise literacy. Among writing instructors, who wouldn't warm to this demonstration of one of our cherished beliefs? Moreover, we must recognize that faculty trained in writing or literary studies will certainly privilege the kinds of literacy practices (traditional essay writing high among them) in which they themselves were trained. We might be asking too much of people to set aside such a natural tendency and fully embrace the anxiety-inducing "unpredictable directions" of multimedia composing. At the same time, we must note how the video and its unrelenting linearity, its inevitable conclusion, rob it of a fuller explication and exploration of literacy. The possibilities for a rich discussion of literacy, of multimedia literacy, seem missing here, even as some technical prowess is demonstrated. There's an argument, for sure. But beyond that?

Two issues are important here. One, the videos thrill us with technical prowess, with the demonstration of abilities that seem so beyond the capability of many of us who grew up before the widespread accessibility of computers and digital video. We are tempted to praise and extol the technical while overlooking the critical and rhetorical shortcomings. Such projects look good; but what do they really say? Second, we worry that such projects work more like linear essays *on* literacy than as videos *about* literacy. It seems to us that we, as compositionists, have concerned ourselves quite a bit with how we can replicate in the new media some of the more traditional ways of storytelling, of explicating, of arguing that characterize traditional texts—texts that we ourselves have critiqued.

In many ways, the commonalities and traditional textual characteristics we have seen in these videos should not surprise us. Since they are produced in composition courses, their general emphasis on text and linear argument makes sense given what we know, not just about the privileging of particular communication modalities in comp courses but about genre theory as well. In *Genre and the Invention of the Writer*, Anis Bawarshi argues compellingly that

> the writing prompt does not merely provide students with a set of instructions. Rather, it organizes and generates the discursive and ideological conditions which students take up and recontextualize as they write essays. As such, it habituates students into the subjectivities they are asked to assume

> as well as enact—the subjectivities required to explore their subjects. (144)

We believe the same generation of discursive and ideological conditions occurs when students compose not just essays but *video* essays or projects as well. The call to compose a video literacy narrative, particularly in the context of the writing course, often situates students to inhabit subjectivities that value textual production. We can see this subjectification at work, for instance, in Eric Wooten's video. At the beginning of it, he wonders if he will ever find anything good to read. By the end, he has found enjoyable reading material: cue triumphal music and two thumbs up. This video traces the performance of a subjectivity—the journey to reading enjoyment—that is in many ways evoked and even mandated by the course context in which the video is produced. What literacy instructor doesn't want to see his or her students come to enjoy reading? So, even though we do not know the specific assignment to which Wooten responds, we can nonetheless surmise that the course context conditions to no small extent the kind of subjectivity that he performs for us in and through the genre of the video literacy narrative. Put another way, Wooten may feel compelled, as Bawarshi might put it, to invent a writerly subjectivity, even as he works within the genre of the video narrative.

How might we use genre to invite students to experiment with and expand the range of subjectivities they can inhabit and perform? Bawarshi argues that "by expanding the sphere of agency in which the writer participates, we in composition studies can offer both a richer view of the writer as well as a more comprehensive account of how and why writers make the choices they do" (144). Again, our interest is in inviting "writers" in our classes to conceive not only of writing in more expansive ways, but *composing* (more generally defined) as well. We wonder what it would be like to offer a "richer view" of the composer by providing a "more comprehensive account" of the rhetorical capabilities not just of textuality but also of multimodality. Such a view requires that we look more closely at genre, and perhaps that we expand models for how students engage and play with genre, so that textual practices don't necessarily dominate multimodal composing projects. In terms of video, for instance, if we teach video narratives in the composition class, we should consider more specifically the particular rhetorical affordances of video so that students can encounter the genre of the video narrative and inhabit it with and perform subjectivities that might exceed the textual. Bawarshi suggests that "teaching invention as a process in which writers access and locate themselves critically within genres" can enrich students' experiences of writing in particular and communication literacy in general (144). We must reenvision the processes through which we teach the genres in which we ask students to compose so that they have a strong sense of the possibilities—and so they do not (and we do not ask them to) transport the values of one genre or medium into another.

COMPOSING VIDEO AS RE-VISIONING THE ESSAY

Such reenvisioning requires a more critical understanding of what is "new" (and what is not) about "new media." As we have suggested, many in our field seem somewhat dazzled by the "newness" of new media, and being dazzled makes us less likely to see critically what new media offer us in terms of composing power. It can, more pointedly, prevent us from seeing that new media, in fact, have a history—complicated, contradictory, ultimately unknowable mixes of history. The technologies that form the constellation "new media" are never innocent and carry with them a reach of ideological DNA that exceeds our own grasp. Thus, when we attempt to fold new media into the genealogy of writing technologies, or the history of rhetoric (now "visual rhetoric," now "aural rhetoric," now "digital rhetoric"), or when we attempt to splice them into our own discussions of business and/or technical and/or professional writing, we necessarily leave out some other equally possible progenitors of new media. Put another way, we have been so taken with the concept of the new media as "new" that we forget that new media bring their own histories with them. Further, as Mary E. Hocks and Michelle R. Kendrick argue, to overlook the "dynamic interplay that *already exists and has always existed* between visual and verbal texts" is to "overlook insights concerning that interplay that new media theories and practices can foster" (1; emphasis added).

Certainly, some new media texts proceed as a set of rational arguments and well-reasoned positions. Just as certainly, there are any number of other histories that offer us insight into the productive experiments and excesses of new media. For example, Jonathan has been researching and writing about the rich history of twentieth-century avant-garde and experimental film, questioning whether students' understanding and practice of communicative action and rhetorical possibility could be enhanced—even altered—if they drew inspiration from a study of early experimental film. Jackie has been researching the written and visual work of French artistic movements of the 1960s to do much the same, and has also been using the irrational, surreal "performative documentaries" of Lourdes Portillo to push her students to a critical engagement with multigenred work.

Indeed, the basic components of the "new media"—technological innovation in dissemination, the use of multiple media, the mixing and remixing of content, and the awareness of the technological medium as intimately connected to the "content," if not actually inseparable from it—all of these have characterized the last 100 years of avant-garde film as well as the art/graphic-design-as-protest movements in the last several decades and the experiments of contemporary filmmakers like Portillo. From the early media experiments of the Dadaists and Surrealists, to the rise of experimental film, to the complex and provocative media games of the Situationists, to the explosion of media-savvy pop art, the avant-garde has taught us that, in a media-saturated society, playing with multiple media in new and challenging ways is a necessary condition for (1) approaching media-overloaded audiences often dulled by media saturation and (2) expanding the rhetorical horizon of possibilities for meaning-making and critical engagement.

Let's play with the idea of possible histories. Students believe they are, in general, film savvy, and the language of movies is part of their cultural vocabulary. However, it is often a historically unrooted vocabulary, an approach to film free-floating in the present.

To challenge this presentism, we ask our students to consider the development of special effects in film, beginning with the groundbreaking work of Jean Cocteau in *La belle et la bête.* The candelabras held by hands emerging from walls, the same hands that then point the characters in the directions they need to go, literally and figuratively—such simple devices, easily imitable, were co-opted by Disney in its animated feature *Beauty and the Beast,* yet few know that Cocteau's avant-garde and pseudo-surrealist film was part of the visual inspiration for Disney animators. Cocteau wanted to access the unconscious, and his surreal images are designed to startle, to make us question what we see. We twitch in our seats; the shock shifts us. What do we see? What is real after all? These images thus perform a critical act that works through the irrational, not the rational, through the body and its sensations, not just the mind. Fifty years later, the handheld candelabra becomes the dancing candlesticks, part of the clever schtick of the Disney film, but presented in less provocative ways. Questions emerge. How do visual effects become co-opted? How do they become commonplaces? How do they change in meaning-making capacity and density over time, losing some meanings but gaining others? How do mediated images lose bodily impact? And what do we gain in recovering a sense of the history of such images?

Jackie approaches such questions by having her students analyze the work of Portillo, Mexican American documentarian and video installation artist, whose films engage difficult and complex political issues such as the ongoing murders of the *maquiladoras* and other young women in Ciudad Juárez. In Jackie's English 240 ("Writing in the Public Sphere") course at Cal State San Bernardino, for example, she assigns the following critical response:

> According to http://www.lourdesportillo.com, Portillo's film *El Diablo Nunca Duerme (The Devil Never Sleeps)* "mines the complicated intersections of analysis and autobiography, evidence and hypothesis, even melodrama and police procedure." Her film *Señorita Extraviada (Missing Young Woman)* offers a similar multilayered approach to explore *what hap-*

> *pened*. In both films, Portillo looks at both "rational" and "irrational" explanations of tragedy, using different genres—autobiography, police "mystery" shows, poetry, *telenovelas*, and others—to present the "truth" of the situation.
>
> As Rosalinda Fregoso writes, "[making] a film about an event that is ongoing and continues to unfold is an inherently challenging undertaking" (25). As we discussed in class, finding and/or writing about the "truth" of ongoing events is also challenging—and it's this challenge that disrupts Habermas' idea of a "rational" and "logical" public sphere. For your response, I'd like you to answer one of the following questions:
>
> 1. *To what extent can you see the "complicated intersections" of genre and the "irrational/logical" explanations play out in the public sphere, especially as that sphere "works" online (in blogs, websites, news sites, etc.)?* or
> 2. *How might we productively and deliberately use different genres (and rational/irrational explanations) online to help us portray the "truth" of an ongoing situation?*

By expanding our sense of the history and context of mediated images, we gain a greater capacity to "read" those images in culturally and even politically significant ways. Yet we also recover a sense of how "texts" do a lot of different kinds of work—how they make meaning and argue in ways that defy some of our more tried-and-true compositional techniques. We recover a sense that the new media understand critical engagement in complex and sometimes nonrational ways that implicate subjectivity and the body in meaning-making.

To show how such a rehistoricized or more richly contextualized approach to new media might enhance students' production of video projects, particularly the video literacy narrative, we turn now to two quick case studies of experimental courses that have attempted to mount multimodal assignments in course contexts focusing specifically on the rhetorical affordances of new media. Both were identified, on their home campus, as "writing" or "composition" courses; and in both cases, the instructors consciously attempted to shift the conceptual frameworks of the comp course so that the traditions and histories of multimodal composing were privileged. At least initially, the courses focused first on histories of media, particularly video media, before inviting students to situate themselves as literate subjects in their work.

The first course comes from a lower-division writing sequence in which students, in a first-quarter composition course, focus on developing their ability to think rhetorically. The course, called "39B: Critical Reading and Rhetoric," is writing-intensive, with students contributing nearly daily to a wide variety of online and in-class writing forums. Numerous short, low-stakes assignments focus on issues of genre, audience expectation, rhetorical strategies, and critical reading. The first major writing assignment consists of a "rhetorical analysis," in which students identify the key rhetorical strategies at play in one of the primary course texts. A concluding assignment, called "Rhetoric in Practice" or RIP, asks students to compose a project (note: not necessarily a *text-driven* piece) that shows students' ability to put some rhetorical strategies to use. The project also requires an accompanying memo in which students reflect critically on the rhetorical strategies they used.

At the time of our work on this book, the RIP assignment, which accounts for 30 percent of students' final grade, reads as follows:

> For this project, you will determine the rhetorical situation of your text, and write a text to fit that situation (the RIP). Then you will write an essay that both narrates your creative and revision process and analyzes the rhetorical choices you made (RIP Essay). Together these two texts will comprise a minimum of 6–7 pages of writing.
>
> Your project must address the class theme and must be written for a "real life" audience. Drafts, peer review and revision are required elements of the assignment. . . . Multiple drafts, peer review and revision are required elements of the assignment, and are calculated into the final grade; failure to complete parts of the process will result in a lower grade. The Project and Essay together must be 6–7 pages long and be

> presented in MLA format, including Works Cited. A total of three (3) sources must be used to develop the essay. An Annotated Bibliography may be required as part of the final draft.
>
> Suggestions for RIP projects:
>
> - Article for a specific audience/publication;
> - Fiction or poetry;
> - Remixed written text for a new purpose or audience;
> - Review for a specific publication (book, film, game, restaurant);
> - Speech or presentation (with speaking notes) for a specific audience/purpose;
> - Sound Essay
>
> Your instructor may change the parameters of this assignment by limiting the project to a particular genre or form; please ask your instructor for more details. Expect to present your final work to the class during week 10. ("Assignments")

As we can see, the project is unique for a "writing" course in that the final "text" need not be an actual "text." Performances are encouraged, as well as collaborations among students. In the five years this assignment has been required in 39B, course directors have steadily encouraged multimodal compositions, and many students work with sound and video to complete this project. And while, initially at least, faculty had a hard time determining the "quality" of projects and grading them, enough projects have been completed, examined, and discussed that faculty feel increasingly confident assessing students' ability to demonstrate their use of rhetorical strategies, even if some of the projects might not be successful outside of the course context. Certainly, much of the grade for the project rests on students' abilities to discuss in the memo how and why they used particular strategies; after all, not many writing faculty feel comfortable grading images or oral performances, much less rap songs and videos. Moreover, these courses are not designed to teach those particular kinds of genres or media. Interestingly, however, some faculty have understood the RIP project as the central component of a course whose stated goal is "to teach you how to approach a variety of texts rhetorically and critically," and that asks students to "compose in a variety of genres (essays, presentations, blogs, wikis, and more) for different audiences"—all as "practice [that] will increase your rhetorical know-how" (eee .uci.edu/programs/comp/39B/index.htm). Given such a mandate, the shift in some instructors' pedagogy toward reconfiguring the course as one in *rhetoric* as opposed to *writing* allows them to focus students' attention on particular modalities of rhetorical practice that need not necessarily be textual. Faculty, who include full-time lecturers as well as teaching assistants, are prompted to experiment along these lines since they are allowed to "thematize" their courses around particular figures, concepts, or issues. A local course culture that appreciates popular culture, a culture encouraged and even fostered by the course director, has created many possibilities for sections of 39B to focus on contemporary visual culture.

Not all sections of 39B are so focused on extratextual figures and modalities. Many direct students' attention and efforts to much more traditional textual production. And even sections that take popular and visual culture as their primary objects of study (and even production) contain a great deal of weekly, even daily, writing assignments, as well as the aforementioned rhetorical analysis essay. However, by examining work produced in a 39B course that takes the rhetorical capabilities of video seriously, we can see how some instructors have used their course contexts to open up possibilities for embracing nontextual rhetorical traditions and affordances.

One such section, led by Kat Eason, focused on the figure of the zombie in contemporary popular culture. Students read several short studies and articles, from both popular journalism and scholarly presses, about the zombie figure, but the majority of the course focused on the history of the representation of the zombie in movies and other visual culture. Students compared the zombie as it appears in George Romero's 1968 *Night of the Living Dead* with zombies in more recent films such as *28 Days Later* and *Shaun of the Dead*. Students considered the particular historical contexts in which the movies were made, such as Romero's staging of a black protagonist in 1968 pitted against white, rural zombies seeking to kill him. The instructor also had students examine other aspects of visual culture, such as the use of zombie images in a variety of computer and console games, including *Dead Island.* Students examined the different permutations of the zombie figure and how different cultural producers deploy it rhetorically to accomplish different ends—including making political messages and other ideological interventions in popular culture.

Given the ultimate "product" required of 39B—the Rhetoric in Practice project—savvy instructors understand that such a course is not just a course about a particular theme, but also a course about the *mediation* of that theme. They focus students' attention, for instance, not just on the figure of the zombie, but also on how different cultural producers manipulate images, sound, video, and text to create rhetorical effects. For instance, in Eason's class, students responded weekly to various blog prompts that asked them to consider the historical, rhetorical, and media features of the zombie films students were viewing. Here is a blog prompt from early in the course:

> Granted: Dawn of the Dead is a classic zombie film and all, but it's also totally 1978. No, I don't mean the cheesy latex and way too bright blood. . . . I mean in rhetorical situation. It is a creature of its time, however well it's endured. So, if Magistrale is correct, and the "best horror fiction must be viewed as contemporary social satire that reveals—and often critiques—the collective cultural fears and personal anxieties of everyday life," then what does Dawn of the Dead tell us about the state of the USA in 1978? Do you think we have the same "cultural fears" and "personal anxieties" today? What's changed? What hasn't? Does that affect the impact Dawn has on us? (Eason, "Sample")

Such a prompt focuses students' attention on the historical context in which a particular subgenre of horror film, the zombie flick, emerges and gains a viewership. Eason then moved students quickly to compare films, fine-tuning their understanding of the vexed question of just what constitutes a particular genre.

In a subsequent blog prompt, Eason asked for the following comparative analysis:

The Zombie Values Project

Purpose: to practice rhetorical analysis on a very small part of a larger piece of work. To identify and explain the rhetorical devices used in the excerpt, and relate them to your larger rhetorical goal. To focus on specifics and identify the most important evidence to use in an argument to convince an audience (your peers) that you're right. To practice the fine art of persuasion. And of course, to get practice with that whole oral presentation thing.

The Meat and Bones

As Simon Pegg's article illustrates, there is some, ah, disagreement as to whether or not 28 Days Later is a zombie movie or not. Your job is to decide whether or not it is, and argue accordingly. You may use 28 Days and Dawn as your primary sources (and anything else as secondary).

1. Take a stand. Do you agree, or disagree, with Mr. Pegg's assertion about traditional zombie values in his article? Why or why not? In other words: what are the different rhetorical purposes of turbo zombies vs. shuffling zombies? Is one more contextually relevant than the other?
2. Choose the scene from 28 Days Later or Dawn of the Dead which best illustrates your argument.

    - Explain WHAT the director is doing in that scene as it relates to your argument. This is not a recap of the action! (Zombies are slow. Romero shows us slow zombies in this mall scene because he wants us to have time to relate to and identify with the zombies themselves.)
    - Explain HOW the scene is achieving its purpose, in relation to the larger theme of the film, and how the type of zombie contributes to the film's rhetorical purpose. This may include film tech-

niques or other rhetorical strategies. (Romero uses close-ups and tracking shots to build the sympathy of the audience for the zombies.)

3. Your purpose is both *informative* and *persuasive*. Convince your fellow classmates, but do it with evidence, not appeals to emotion and fancy graphics. ("Workshop")

Such a querying—what really makes for a "zombie movie"?—involves more than simply taking stock of fan bickering. Eason asked that students develop skills in argumentation by paying attention to how genres *develop over time*, and how the genre in its evolution plays with viewer expectations. Shifts in expectation can also signal shifts in value, opening up the possibility of interpretive analysis; for instance, what might the faster-moving zombies in *28 Days Later* signal culturally? Pedagogically, the blog prompt puts into circulation textual analysis (the Pegg article) and visual analysis (the requirement to focus on a film scene) with a consideration of genre. And just a week later, Eason asked students to consider a further twist in genre by inviting them to write about *Shaun of the Dead*, examining in particular how the film uses (but also ignores and subverts) the elements of two distinct genres: the horror film and the romantic comedy. What is at stake in making a zombie *comedy*? At what do we laugh? How has the horror of the zombie film been repurposed to create comedy—and to what effects, socially and culturally? Perhaps the issues of racism and race relations alluded to in Romero's *Night of the Living Dead* have been eclipsed, at least in the popular consciousness, by other concerns. Or perhaps we have learned to laugh at ourselves. Still yet, as one student put it, the recent proliferation of zombie films seems indicative of a "post-life" culture, in which we have difficulty imagining a future that isn't about voracious, mindless consumption; hence, our laughter at a film like *Shaun of the Dead* might come from the comedy of discomfort.

We emphasize Eason's historically based approach to the genre of the zombie film because it offers students a robust way to think about how a genre changes over time, meeting the needs and expectations of different audiences. The play of differences with and *within* the medium is generative. Genres mix and remix over time (horror becoming romantic comedy) to create new possibilities for meaning and interpretation, but they also play with their own tropes (shuffling versus turbo zombies) to signal and produce new interpretive possibilities. Such mixing offers students a powerful example for their own play as they turn to the RIP project, in which they compose their own multimodal work. For example, in "A Costly Increase" (www.youtube.com/watch?v=obFMaqdnbo4), one group of Eason's students created a Claymation video about the University of California's tuition increase. While not precisely a literacy narrative, the video nonetheless comments directly on students' concerns about the rising cost of their collegiate education. In the video, unsuspecting students receive news by mail of hikes in tuition, only to discover that the UC Board of Regents seems to be manipulating increases to create fellow zombies—perhaps a comment on the stifling (dare we say, chilling, even deadening) effects of rising tuition on students' aspirations and ambitions. The Regents' plan, however, backfires as newly zombified students attack the Board of Regents, and the video ends in a typical zombie apocalypse. The music score shifts from pathos to jaunty jingles, signaling a mix of emotions—despair, anger, delight in revenge—and the overall effect is satiric. The video might ultimately simply represent the wish-fulfilling revenge fantasies of students tired of tuition increases, but as such it also gestures toward a pressing fiscal commentary: in an increasingly interconnected economy, passing the bill on to others is bound to rebound, in some way, at some time.

In terms of media-savvy and multimodal education, we are not sure such a video could have been made had students not had extensive time to consider the particularities of the zombie film genre. Tracing the malleability of the zombie image allowed them to envision possibilities for manipulating that trope in the service of social commentary. Moreover, what is particularly compelling about such work is the deployment of video and media tropes to address *stu-*

dents' concerns—something that Eason and her colleagues actively solicit in their courses. The course highlights the history of such mediation so that video becomes present as an adaptable, evolving medium through which students can inhabit a genre and begin to make it their own. In terms of literacy, these student composers articulate their understanding of their own education, their becoming literate, in a much broader economic and institutional context; therefore, we see these zombie videos as much more complex literacy narratives than the ones discussed earlier. Or, borrowing from Bawarshi: having studied the history of a particular film genre and being invited to play with it, these students inhabit a richer subjectivity that allows them to launch substantive critique, not just replicate a position held by the instructor on the value of being literate.

What happens when an entire course is devoted to examining the history and rhetorical possibilities of new media, particularly as they relate to self-representation and subjectivity? What kinds of "essays" or multimodal projects emerge from such an experience? Elizabeth Losh experimented with just such a course, "Digital Rhetoric: Becoming a Conscious and Critical User of Social Media" (https://eee.uci.edu/08f/25823). Focusing on the history of the development of new media, particularly what is now called "social media," as well as on accessible theoretical work analyzing new media, Losh developed a course in which students read about but also developed and designed their own social media sites, including blogs, wikis, personal websites, Second Life avatars, and YouTube videos. Pitched as an upper-division writing course, "Digital Rhetoric" was "designed to make students more effective creators of social media and to give them [a] more theoretical perspective about the conventions of online communication"; to facilitate such a goal, Losh encouraged students to be "active content-creators of curricular online materials." Her "Reading and Viewing Assignments" were broad-ranging but attempted to give students a sense of the historical development of thinking and theorizing about new media. The course culminated in a YouTube video project, for which students were asked to compose a video:

> Write a proposal of 200–500 words that describes your YouTube video essay, which must be on a topic related to digital communication or social media that is relevant to the issues raised by the course. Explain what other videos are available on YouTube on similar subjects and list at least one possible YouTube video that you might choose to respond to in order to get your video viewed by members of the general public.
>
> Think about if you want to use images, footage from creative commons sources, or fresh footage that you have shot yourself. Give the reader information about your argument, the central claims, and the evidence that you plan to present. As you work on the script and editing, you may find that this basic proposal needs to be revised, but it is important to start with a sense of rhetorical context, audience, and purpose. (Losh)

Losh staged the assignment so that students not only made an initial pitch, grounded as a possible response to another video, but also submitted images, a script, and video "drafts" for instructor and peer review. As a whole, the course context of "Digital Rhetoric" provided a substantially rich environment in which students learned about media and also participated actively in a wide range of social media while reflecting on that participation and putting their reflections into conversation with their emerging sense of the development of media over time. The advantage of thinking across multiple media is clear: students could see the different rhetorical possibilities afforded by different media, weighing what a blog or wiki could do versus a video or Facebook page. Losh's final assignment then asked students specifically to put into practice and reflect on their own emerging media literacies—or, as the course subtitle puts it, to think about how they were becoming "conscious and critical users of social media." Creating a video for YouTube seems an appropriate culminating experience in that video combines image, sound, and text—a robust multimodal project. Since students were asked to consider their video as a potential response to another video, they had the opportunity to further, critique, or comment on an existing conversation about social media.

We can catch a sense of the richness of this commentary in a video essay composed for this course. This video, "I.D. / self :: the new 'real'" (www.youtube.com/watch?v=1BhEj-tI66E) challenges us to rethink what being "literate" means, particularly in terms of our engagement with new media ecologies (newartisticdirection). On the surface, "I.D. / self :: the new 'real'" is about one student's ("Johnny's") engagement with the new media, particularly Facebook, AIM, and *World of Warcraft* (*WoW*). After an introduction that raises some intriguing questions about identity on the Internet, the video proceeds with a clever explication of the three major domains through which the student engages others via the Internet and various multimedia platforms. Throughout, the composer uses the visual cues associated with different platforms—the blue bands of Facebook, the chat bubbles of AIM, the gothic scripts of WoW—to signal what self is on display. A conclusion, though, raises more questions than it answers. By the end of the video, we are unsure who the "speaker" here is. Who is "Johnny"? A high school student? A college student? Is he a he, or perhaps a she? The point of the video, despite its apparently linear format (intro, three examples, conclusion), ultimately seems to be to raise even more questions about the possibility of knowability in digital spheres than it sets up to address in its introduction. The video is not about clarifying, being rational, revealing "truth." It's about reflecting on the media itself and the process of subjectivity becoming mediated. We are reminded, when watching this video, of Magritte's *Ceci n'est pas une pipe*. This is not a pipe. Essentially, the video says, *This is not Johnny*. We see bits and pieces of the whole, but never the whole—which perhaps doesn't even exist. In the contradiction, in the fantasy, in the elisions—we see and don't see; truth is both revealed and hidden. Indeed, what is perhaps most startling about this video is how it uses the temporal linearity of the video medium, unfolding as images in time, to double back on itself, so that, as the video proceeds, you steadily question everything you've seen.

We might argue that this video performs *metonymically*. That is, its constituent elements—from the opening and concluding framing shots to the different visual presentations of self—constitute pieces that allude to an absent whole. This video suggests to us a new way of configuring the relationship between compositional production and the audiences such production serves. Composition, in its traditional forms and genres, often privileges the metaphor over the metonym and thus often works against the metonymic possibilities of some new media. For instance, in many of our courses, we strive to have students develop arguments that are "like" full-scale debates, rational encounters in the public sphere—that argue through points as though there were different interlocutors hashing out an issue, and as though resolution were potentially possible. Wooten comes to value literacy through a journey, with a beginning, middle, and end, metaphorically mimicking the move from ignorance to knowledge. In contrast, a metonymical approach to argument offers us the bits and pieces but never pretends that an understanding of the whole is either possible or desirable. Johnny's video shows us a "whole," a "totality," a "real" identity that finds metonymical figuration in the various "cases"—but the case studies, proceeding as a mimicked five-paragraph theme, empty themselves of meaning by the end of the video. We are led to question self-revelation itself: "they reveal something about me . . . don't they?" Johnny's various cases, assertions, and bits and pieces of evidence invite us, ultimately, to question what we know—and that questioning proceeds more through surprise, contradiction, and fantasy than through debate, argument, and reason. The video also points out what is finally incommensurable in our understanding: it embraces contradiction and irresolution as powerful ways of knowing.

Both Eason's and Losh's courses push the boundaries of what composition's engagement with new media can—and probably should—look like, even as they enact two different approaches. Eason's course layers a rich rhetorical understanding of new media into a thematic focus on a powerful figure—the zombie—that moves across multiple media. Losh's course works directly with conceptualizations of new media rhetoricality, taking the authoring of new media projects as the thematic content of the course itself. In both cases, composing has become radically remediated away from the primacy of written and print-based texts and toward the inherent multimodality of communication in contemporary public spheres.

We can see such multimodality at work in the public sphere in another video, this one not produced for a course but intended nonetheless to be pedagogical in a broad sense, and one that attempts to expand our sense of the "literate," or how one can speak publicly about issues of personal, and ultimately political, import: LaReina DelBarrio's response to Barack Obama's decision to have Rick Warren offer the invocation at his first presidential inauguration (www.youtube.com/watch?v=bgXpMVpAZfA). DelBarrio's voice-over slides in and around a variety of provocative images—flags waving, Obama smiling, and drag queens dancing. The video performs simultaneously a sense of queer eroticism and queer outrage about Obama's choice of Warren, who had been pivotally involved in the 2008 overturning of marriage rights for California gays and lesbians.

Note, though, that there is little direct argument here: the video is more about mood. It is about pain, resentment, and other emotions that cannot be squared against political realities; it is also about those political realities that reveal themselves to be compromised, even contradictory. The images speak to this incommensurability in the juxtaposition of the drag queen, the little red schoolhouse, and the pervasive American flag flapping in the background. The embracing gestures of the drag queen collide with the static images of Obama, while the voice-over narration and the text on the screen speak to each other of outrage, shock, and disbelief. More provocatively, the eroticism of the drag queen, slightly bent over as she wiggles her ass at Obama, suggests both a "desire" for recognition, for an intimate embrace into the public sphere, and a sense that we have been bent over and fucked once again. At the end of the video, the drag queen covers her eyes, mimicking both the sense that justice should be blind and that, in this particular case, the reality of injustice is too difficult to witness. Granted, this is hardly the traditional literacy narrative, but we include it as an appropriate counter to the foregoing video literacy narratives because it offers a complex set of textual, visual, and multimodal statements on the difficulties of finding pro-queer expression—of being legibly literate—in a society embracing homophobic and anti-queer policies and positions. In other words, it uses multimodality to challenge and hopefully expand our sense of the literate, and of the struggle some face to be understood and heard in the public sphere.

We wonder, as an imaginative exercise, what these videos would look like if they were rendered to us as "proper essays." What we would find are most likely essays that would (1) diminish or even potentially resolve the contradictions offered in the videos in favor of an agreeable, Rogerian-esque compromise ("We disagree but are not disagreeable") or (2) perplex us with textual experimentation that would not look like a traditional essay, that would proceed more through montage, rant, diatribe, and less through reasoned debate. How, for instance, might Johnny's video not work if it were an essay—either a five-paragraph essay (that it mimics) or a more aesthetic, Ezra Pound–like montage? These imagined essays might rationalize the various issues they raise. We might, for instance, read an essay that offers reasons for why Obama did what he did. But those reasons wouldn't do justice, we believe, to the necessary outrage, even the necessarily perplexing and perplexed eroticism of DelBarrio's video. A rationalizing argument can't quite encompass the embodied sense of disappointment, hurt, rage, and even disgust that is a crucial part of understanding a queer response to Obama's politicking.

SOME FINAL CUTS

Our examination of video in this chapter allows us to consider how we might question the legitimizing moves of our discipline, the ways it attempts to make itself whole and sound, and how certain histories, certain excesses, and certain compositional possibilities are left out. We believe that remixing histories, even (or especially) those excessive histories, is a starting point for interacting more generously with new media—not to colonize, or subsume, or contain, but to celebrate composing in all its multiple potentialities.

Certainly, not all students may be able to compose the kinds of pieces we have lauded here. We must ask, however, how might students' critical vocabularies of media gain in sophistication if they can trace the possible histories of new media, of technology, of composing? And how might experimentation with media practices allow them access to ways of knowing that are not tied to rational exposition or narrative development? Such explorations of alternative, nonrationally developed kinds of arguments may enrich their understanding of how arguments are made, how textual and visual material can function rhetorically in ways beyond logos, how affect is rhetorically usable, and how such communicative action can be a critical enterprise that often tests the boundaries of intelligibility—of what is known and knowable. If students have the chance, even in small ways, to see how a Jean Cocteau or a Lourdes Portillo or a LaReina DelBarrio expand the "grammar" of film, then they have the chance to become more literate consumers of video, and perhaps more sophisticated prosumers as well. At the very least, what we offer them is a chance to think beyond the formula—the narrative exposition, the developed rational argument, even the clever parody—to explore possibilities of textual, visual, and multimodal production that could be rhetorically richer.

Along such lines, we return briefly to the video discussed at the very beginning of this chapter, Kyle Kim's "Closer." In the instructor's reflection on his student's work, he comments that Kim shows us a picture at one point of Kesey's *One Flew over the Cuckoo's Nest*, noting that this is one of the few substantive moments in which Kim uses a "textual" bit of meaning-making in his film. The instructor clocks this moment, having hoped for more of them, but appreciating it nonetheless. We might argue, though, that the instructor has missed a greater point. Kim's use of Kesey's novel—thematically appropriate, as the instructor notes—is also *historically* rich and suggestive. For just as Kesey creates a first-person avant-gardist narrative to critique mid-twentieth-century notions of self-awareness and intimacy, so too does Kim, on a smaller scale, perform likewise for his generation, using one of its favored modalities—the music video. An argument is being made here, but it is one that proceeds by valuing and forwarding not rational and linear argument but iteration, citation, association, and references to the *irrational*. It also references its own indebtedness to traditions of experimental aesthetic production, even if these are largely textual. In this way, "Closer" riffs on the genre of the literacy narrative; its studied use of multiple literacies is precisely what makes it so interesting as a video—and so interesting as a "text" that forwards its maker's awareness of the avant-gardist traditions in which he composes.

Ultimately, we might suggest that we need to reconsider more critically our disciplinary divides, particularly those that we use to legitimate ourselves. We understand the importance of authoring "composed" essays; we don't deny the very real and material need to help students develop the kinds of compositions—the kind of composure—that make them legible in the marketplace, not just of ideas but of hard currency. We also do not deny the necessity of composing such texts ourselves (more on this in Chapter 3). But we want to also make room for the kind of "writing"—and the kind of subjects—that challenges such composure, that offers rich and (yes) excessive ways of thinking and writing. We hope that our critique in this chapter allows you to consider how we might collectively, as a discipline, question the legitimizing moves of our discipline, the ways it attempts to make itself whole and sound, and how certain histories, certain excesses, and certain compositional possibilities are left out.

3

Prosumerism, Photo Manipulation, and Queer Spectacle

> "There has always been more to the image than meets the eye."
> —David Blakesley and Collin Brooke

ONE OF THE KEY RHETORICAL AFFORDANCES of new media in general is the capacity for active writerly participation in complex public spheres. As Kelly Kinney, Thomas Girshin, and Barrett Bowlin write, this is an important signifier of the "third turn to the social":

> [While] "the social turn" represents many things to many people, we see three distinct shifts in this so-called turn. The first . . . emphasizes teaching writing and learning how to write as collaborative, interactive processes. The second shift grows out of the first, but, rather than focusing primarily on instructional practice, as James Berlin writes in *Rhetorics, Poetics, and Cultures,* it examines and critiques the signifying practices that shape subject formation—and, by extension, the discipline—"within the framework of economic, social, and political conditions" (83). While scholarship represented by the third social turn does not ignore classroom pedagogy or critical theory, it also does something quite more: it takes as its starting point embodied activism.

While Kinney and colleagues do not point specifically to new media as a playground for embodied activism, it is not too difficult to find a resonance here with social media actions; note, for example, the Twitter feeds that fed the Occupy Wall Street movement, or the large-scale Facebook and Twitter uproar that resulted from the

Susan G. Komen foundation's decision to pull funding from Planned Parenthood. Indeed, this sort of action within new media points to what Daniel Anderson has called "prosumerism." In "Prosumer Approaches to New Media Composition: Consumption and Production in Continuum," Anderson argues that prosumer approaches to literacy encourage not just the analysis and consumption of texts (particularly new media texts) but also the critical production of such texts. The prosumer, he says, refers to a convergence of the professional and the consumer in terms of new media tools; this convergence "allows professionals to do things with less, and those who have less, to do things that professionals have been in the past only able to do." Further, such a convergence creates the conditions for more active, critical consumption and production in the public sphere, particularly in Warner's "poetic" sense of public discourse. As noted in Chapter 1, this connection between multimodal literacy and civic action has increasingly become part of our scholarly discussions.

This emphasis on not only consumption but also critical production of texts is important to new media literacies for it points us away from the techno-inclusionism that always sees technology in service to established composition principles. For example, books such as Wysocki et al.'s *Writing New Media: Theory and Applications for Expanding the Teaching of Composition* (2004) and Ball and Kalmbach's *RAW: (Reading and Writing) New Media* (2010) focus on the critical, engaged production of new media texts. Even earlier, Mary Hocks spelled out a pedagogical emphasis on analysis and production in her 2003 "Understanding Visual Rhetoric in Digital Writing Environments." Hocks spends the first part of her essay offering a framework for analysis of online texts, including such criteria as audience stance, transparency, and hybridity. Critical to our purposes here, however, the second half of her article discusses not analysis of online texts, but production of those texts. Hocks writes, on teaching visual digital rhetoric:

> When we bring an understanding of digital rhetoric to our electronic classrooms, we need to expand our approach not only to rhetorical criticism but also to text production. Digi-

> tal technologies can encourage what the New London School theorists call a multimodal approach to literacy, where using communication technologies engages students in a multisensory experience and active construction of knowledge. (644)

Drawing on the work of Gunther Kress, Hocks distinguishes between critique and design, writing that to have a "balanced" approach, "we must offer students experiences both in the analytic process of critique, which scrutinizes conventional expectations and power relations, and in the transformative process of design, which can change power relations by creating a new vision of knowledge" (644–45).

This critically informed production, like Anderson's prosumer approach, and like the "reflective praxis" at the heart of Stuart Selber's approach to "multiliteracies," forms a key component of new media literacies, whether those literacies take the networked form of a website or a blog, or the nonnetworked form of a manipulated photo. In any case, as Cindy Selfe points out in an interview with Brian Bailie, if the medium isn't the message, it still forms an important part of it:

> But media themselves can change rhetorical approaches. Media, for example, aren't transparent or neutral, so we need to practice with media, and we need to know the affordances and the capabilities and the tendencies and the ways in which particular media and particular modalities shape our expression, before we can be rhetorically effective. If we don't do those things well, it doesn't matter how important your rhetorical purpose is or how focused your rhetorical intent is or how keen your rhetorical understanding is, you have to know how to work with the tools. Unless you know something about the tool, you're not going to be as effective in deploying the rhetorical affordances of that tool expertly. (Bailie)

Yet even while new media scholars call for increased attention to production, a good deal of composition studies scholarship done on visual rhetoric remains analytical rather than productive. Dànielle Nicole DeVoss and Julie Platt comment in their online essay,

"Image Manipulation and Ethics in a Digital-Visual World," on their own focus on critique rather than production: "Although rhetoric and composition has a long, rich history of deeply interrogating texts (both alphanumeric and image-based), we haven't particularly honed in on the ways in which digital tools allow for the easy manipulation of images, or attended to the ethical implications of such manipulation."

This fear of manipulation—of modern sophistry, if you will—still hamstrings much scholarship on new media and visual rhetoric, particularly if that scholarship is print scholarship. Again, there are notable exceptions, particularly the work of Wysocki, the "hybrid essays" of Myka Vielstimmig (Kathleen Blake Yancey and Michael Spooner), and much of *RAW* (Ball and Kalmbach). However, these are notable exceptions. Perhaps our too-easy assumption that "new" media means "nonprint" media feeds this fear and the concomitant reluctance to engage with multimodal tools; perhaps not. But certainly it points to a techno-inclusionist emphasis on essayistic literacy, particularly our own as scholars and teachers of writing. As noted in the first chapter, Carolyn Handa's introduction to *Visual Rhetoric in a Digital World: A Critical Sourcebook* reassures compositionists of their "crucial" knowledge:

> Here is where our experience as composition teachers becomes crucial. Preparing students to communicate in the digital world using a full range of rhetorical skills will enable them to analyze and critique both the technological tools and the multimodal texts produced with those tools. (3)

The implication is that we do not possess the technical skill but we do understand "the importance of thinking carefully about rhetorical questions." Such crucial knowledge of analysis and critique may be part of the compositionist's tool belt, but, as Selfe says (quoting engineers at Michigan Tech), "If the only tool you have on your tool belt is a hammer, every problem looks like a nail." The overemphasis on analysis and critique—well within our compositionist comfort level—can blind us to the critical capabilities of prosumerist rhetoric for ourselves. Further, says Selfe:

> There are still a lot of humanists, who use technology, but don't think about focusing on it in their classes—especially in terms of critically informed production. So while these folks use a cell phone and use scholarly databases and use a lot of websites, and use technology in their classes in terms of making multimodal texts available for consumption by students, teaching students to analyze and criticize mediated texts, I still know plenty of teachers who avoid teaching students how to compose or produce such texts because they personally don't feel it's their responsibility to compose, or to teach composition, in any modality except the alphabetic. (Bailie)

Clearly, critique and analysis are crucial parts of responsible, ethical agency; we wonder, however, to what extent scholarly reluctance to engage multimodally has foregrounded analysis at the expense of production. This foregrounding of analysis in our scholarship must play out equally in our teaching.

In this chapter, then, we argue that it is crucial for scholars themselves to analyze and compose multimodally, to engage in the critical production of knowledge in the poetic public sphere of our field along with our students. Specifically, this chapter argues for the power of images and mashups to create rhetorical possibilities that not only question standard narratives, but also communicate powerfully in ways that a composed text cannot. Thus, photo manipulation becomes more than invention, a step toward the composed text, but a form of communication in its own right—one that allows us (and our students) to intervene in important and pressing public discussions. We argue, with Sid Dobrin, that visual texts and photo mashups are more than "green" texts, or "texts thought to be immature, undeveloped, or designed for inexperienced readers" (269). They are, because of the turn from page to screen, postliterate rather than preliterate forms of communication (Dobrin 269).

Further, we argue against DeVoss and Platt, who write that the critical production of photo manipulations is merely the recognition of "the spaces where visual manipulation is useful for critique or parody." DeVoss and Platt take a dim view of image manipulation as critical production, focusing instead on the ethics of simulation:

> Historically speaking, image manipulation is certainly nothing new. However, the accessibility of graphic-design tools and the spaces and places for distributing images are relatively new. Further, the tools and the spaces place significant expectations upon us as viewers/readers and producers/creators. Culturally speaking, although we generally know better, our culture tends to cling to a sense of photographic truth and machine objectivity. New technologies allow us to represent ourselves—and, more importantly—to represent others in ways that are potentially problematic. Indeed, we currently live in a culture of simulation—where images are created, or where arguments are created through photographic manipulation.

We maintain that, more than being "useful," photo manipulation can be a *generative* rhetoric, a sophistic excess of meaning, a postliterate "critique or parody" that exceeds the viewer's gaze. One recent example of such productive excess is the 2011 Benetton "Unhate" photo campaign, which featured Photoshopped images of world leaders kissing: the pope kissed Grand Imam el-Tayeb; Barack Obama kissed Hugo Chavez; etc. Nestled in layers of global politics and homoeroticism, the pictures provide ample opportunity not only for shock but surely also for critique and analysis. But can they inspire more than critical consumption? That is, can they offer us possibilities for prosumerist photo manipulation ourselves?

In what follows, we make use of a queer *détournement* of Situationist spectacle as it encounters representational technology in order to model a prosumer approach to and generative rhetoric of photo manipulation, specifically around issues of gender and its figuration in public discourse. The Situationists provide a powerful model for publicly engaging and manipulating mass images, and their strategy of *détournement* in particular recasts spectacle in light of different levels of access to such technologies. Coupled with direct action politics and queer incursions into public discourse, this access to the means of representation creates a curious, evocative space in which to act; at the same time as the spectacular invokes our own need and our own alienation, it also creates a workable space for feminist and queer *technai* of self. It is this workable space, we believe, that promises much expanding of our sense of the rhetorical affordances of visual texts, for it acknowledges and makes use of one of the key affordances of photo manipulations: excess.

THE SITUATIONIST SPECTACLE

"Situationism" was a revolutionary aesthetic movement that, as Fletcher Linder notes, concerned itself "primarily with enabling agency through creating conditions of possibility, which in their case meant placing art in everyday contexts in order to foster a sense that life can be artful, playful, enjoyable—something other than what the situationists felt life was at the time, namely a place of work, struggle, despair, and most of all, alienation" (368). Formed in 1957 when a number of European avant-garde groups (including the Lettrists and the Movement for an Imaginist Bauhaus, among others) joined forces, the movement drew, clearly, from Surrealism and Dada, but channeled its interest in the absurd toward more explicitly political ends than those earlier movements. The movement dissolved into transatlantic factionalism in 1972.

For the purposes of this chapter, we play off the Situationists' key modification to classical Marxism: an expanded view of commodity fetishism that indicts consumer capitalism for its trade in images. The Situationists recast consciousness as that which happens at the point of consumption rather than production, and saw *image* as a more accurate term than *commodity* for a discussion of alienation in the late twentieth century. That is, Marx might argue that alienation derives from the commodification of labor, whereby people's "relations of production assume a material shape which is independent of their control and their conscious individual action" (187); Situationists might argue that alienation derives, in this historical moment, from the commodification (and representation) of consumption (and desire). In this cosmos, the *spectacle* looms large; in using the term, we refer to Guy Debord's 1967 manifesto

La Société du Spectacle (The Society of the Spectacle), in which he names spectacle as the force that simultaneously seduces and alienates people in consumer capitalist societies. Debord, a filmmaker who (along with Raoul Vaneigem and Asger Jorn) might be said to have "led" the Situationist movement, writes that spectacle "is not a collection of images, but a social relation among people, mediated by images" (12), a relation that "far from realizing philosophy, . . . philosophizes reality, and turns the material life of everyone into a universe of speculation" (17).

The Situationists argue that, in order to sustain its own motion, consumer capitalism relies on the construction of "pseudo-needs" that increase and sustain consumption. With the advent of modern advertising and increasingly sophisticated marketing, our lack (as consumers, not as workers) is made visible through the spectacular—not just transactions of images, but transactions of images within a context of continuous consumption. Like any addictive seduction, as we accumulate more of our own alienation, we find ourselves ever in need. To play on René Girard's concept of triangular desire, we are at once subject and mediator of our own desire. The image-as-object is not ownable, but exchangeable; its purpose is to fix—temporarily—a sense of self. In brief, then, spectacle is the ever-present reminder of our own need and the avatar of our alienation. It is the fulcrum upon which our desire for being turns.

For Debord and his followers, spectacle existed as the unfortunate and overwhelming residue of consumer capitalism in a world in which individuals lacked access to the means of (re)production and (re)presentation. It is our contention that current do-it-yourself (DIY) trends in technology (such as photo manipulation), combined with an ongoing questioning of sexuality and text, have created a rhetorical moment—paradoxically, within and through the spectacle itself—in which to battle alienation. Yet even in this reiteration (with a twist) of the idea of the spectacle, we engage in a move that Situationists would recognize and even endorse: *détournement,* which is, as Frances Stracey explains, "the Situationists' term for a technique of pillaging and reappropriation" (59). Our project, then, will hark back to the Situationist goal of combating alienation through "popular action to reclaim creative control of everyday life as well as political control of the means of production" (McIver node 2 par. 1).

THE *TECHNE* OF QUEER *DÉTOURNEMENT*

> People who talk about revolution and class struggle without referring explicitly to everyday life, without understanding what is subversive about love and what is positive in the refusal of constraints, such people have a corpse in their mouth.
>
> (Vaneigem 26)

We'll start simply: Jackie put the following pictures—of her head on someone else's body—on her office door. The reactions: laughter; jokes about how Pilates was really working; a number of comments about how hot she was. However, a number of people—good friends, even—were disturbed: "I can't even look at that—it's too creepy!" The queer reading is, perhaps, obvious; in the pictures, there is a slipperiness of gender and representation, a layering of possibility and dissonance, gaps and excesses of meaning. As each reader approaches the text, she or he has different (and contradictory) interpretations of that body and what technology has done with it. In that mesh of possibility is a great deal of discomfort, a discomfiture courted by queer readings; for like other texts, these pictures explode the possibility of a sexuality-revealing "definitional center" for either the material body reading or the representation read. Indeed, as Eve Sedgwick writes, "a lot of the most exciting recent work around 'queer' spins the term outward along dimensions that can't be subsumed under gender and sexuality at all: the ways that race, ethnicity, post-colonial nationality criss-cross with these and other identity-constituting, identity-fracturing discourses, for example" (8–9). Further, she writes, "queer" derives its critical weight from first-person iteration: "A hypothesis worth making explicit: that there are important senses in which 'queer' can signify only when attached to the first person. One possible corollary: that what it takes—all it takes—to make the description 'queer' a true one is the impulsion to use it in the first person." (9)

Yet *queer* is also problematic in its first-person iteration, for it may elide *gay* and *lesbian* as transgressive terms. In "Doing It in Class," Erica Rand talks about the perils and pleasures of using sexually explicit queer images in her twentieth-century art classes. Using these images provides a direct and compelling means to raise issues of power and pedagogy, of censorship and author's voice, of encountering an Other for oneself and coming face to face with one's own ideas of "the obscene" or "the erotic," with the way that "censorship in practice actually functions to reinforce existing power inequities since it is usually deployed by the overempowered against the underempowered" (29; author's emphasis). The illustra-

tion for Rand's article is a line art image of two women wearing T-shirts that say, "Dykes Fuck. Study That." We are offered, too, a line art magnifying glass that enlarges the words as well as the women's smiling faces.

Rand's article is exactly on point—showing how a seemingly transformative pedagogy is stymied by very real considerations of legal and professional consequences, considerations that take into account both feminist and conservative arguments against pornography. However, more than that, we want to note this potential irony: the continual move to tame the queer, to render it familiar, to make it unqueer. "Queer," in fact, once a harsh and potentially life-threatening marker, then an in-your-face invocation of difference (Queer Nation, etc.), can now function, via the neutering force of corporate media—the Situationists' spectacle—as tame, even friendly; just note the friendly folk that were on *Queer Eye for the Straight Guy.* Further, as a number of scholars have noted, the division between LGBT studies and queer theory is a deeply contested one. Linda Garber writes in *Identity Poetics*:

> The divide, in certain ways, is not a divide but rather something like a failure to communicate, coupled with the coercive, divisive pressure of the academy to smash our forerunners—even if it means misreading and/or misrepresenting their work—to establish our own careers, which others will later take apart. . . . The point is not that one [theory] is right and the other wrong, nor that one type of theory is smarter or more sophisticated than the other, but that either taken alone leaves great patches of the theoretical canvas bare. (5, 7)

Keeping this potential divide in mind is key to any first-person iteration of "queer." Yet the tension of that divide is part of what makes first-person queer so potentially powerful. To perform queer first-person is indeed a radical, disruptive, even abnormalized invocation of body, gender, desire, fear, and sensation. It is a spectacular act, in which we might make use of our converging alienations, our mesh of desire and want, in order to position ourselves to be—if only for a particular, rhetorical moment. Even more, because of the constant exchange/deferral of need, queer first-person increases and sustains itself through its desire, serving as the engine of its own perpetual visibility. It is, simply, one act in a generative *techne* of self.

Techne has no precise equivalent in English; it has been variously translated as "art" (or craft), "technical knowledge," "skill," etc. Here we use *techne* as a sort of praxis middle ground, more than the "clever, bold strokes" of *phronesis* (Lyotard and Thebaud) or the knowledge-making systematicity of *episteme*. We pose the *techne* of queer sexuality as a sort of generative lived knowledge; it is a view of *techne* that points less to the prescriptive how-to sense of the term and more to the ethical, civic dimension. This *techne* has two broad parameters: (1) the acknowledgment and even embrace of the idea of spectacle, the alienating distance between bodily self and representation as a productive space for critique; and (2) the importance of lived experiences to the formation of an ethical stance. The life of the body is not to be ignored.

PROMISCUOUS PHOTO MANIPULATION

> We have a world of pleasure to win, and nothing to lose but boredom.
>
> (Vaneigem 10)

What promise might such a *techne* hold for working with images, for photo manipulation as a generative rhetoric? In their 2001 introduction to a special issue of *Enculturation* on visual rhetoric, David Blakesley and Collin Brooke write:

> Any expression in words or images makes assertions that reflect the perspective and attitude of those who name, and (potentially) influences the interpretation of the hearer, the reader, or the spectator. When you consider as well that the process moves in both directions (seeing itself as the expression or assertion of power), you have the invitation to consider the visual and the textual as inextricably linked in the parallel acts of believing, interpreting, persuading, and identifying—each a rhetorical act in its own right worthy of a closer look. (node 4)

Part of this closer look, we believe, is moving beyond analysis and critique, moving beyond composition to embrace what Geoffrey Sirc calls an "art stance to the everyday" that depends on simultaneous analysis and production, as well as production as analysis ("Box-Logic" 117).

In the past few years, we have taken seriously Sirc's call, not just in the design of photo manipulations to shock friends and colleagues, but as a pedagogical tool to provoke critical reflection. One venue for such an emphasis has been Jackie's English 658 course. English 658, as described in Cal State San Bernardino's *Bulletin of Courses*, is a graduate course that explores "the ways in which computer technologies can be integrated into composition courses and literacy contexts of the workplace, and how they alter the understanding, acquisition, and teaching of literacy in our society and culture" (491). This course, then, does not train students in the brass tacks of instructional technology, but rather focuses on contexts and histories of computers in writing. It focuses, that is, on the "why" more than the "how to." At the same time, the course is a "praxis" course, one of several in Cal State San Bernardino's MA in English Composition program that teaches future teachers how to work theory into informed practice. Thus, it can serve as a powerful venue for those future teachers to develop their own prosumer approaches to writing multimodally. The opening writing assignment in the class has been the techno-literacy narrative, a visual essay adapted from Cynthia L. Selfe's "Toward New Media Texts" in Wysocki et al.'s *Writing New Media.* As teachers, Selfe writes, we must "pay more serious attention to the ways in which students are now ordering and making sense of the world through the production and consumption of visual images" (72). For the essay, students reflect on their own *techno-literacies,* broadly defined as critical literacies of multimodality, and then compose a visual essay that conveys their experience as techno-literate people. Such essays have taken the form of websites, Facebook pages, posterboard presentations, iBooks, PowerPoint and Prezi presentations, YouTube videos, and other genres.

Regardless of genre, however, what is key to the assignment is the idea that teachers themselves must compose visually, must question their own literacies, must critique and design multimodally if they are to teach future students well. One way to pay more serious attention is to work with visuals ourselves. We have done so, modeling for our students and others in the field our reflections through photo manipulation of our self-representations, our figurations, and our desires. For instance, riffing on her early photo manipulation door art, Jackie created a series of manipulations that first showed Jonathan in bed with himself, and then Jonathan and Jackie in bed together, each holding a piece of technology (Jackie a television remote, Jonathan a laptop). The images parody a romantic couple in bed together at the end of a long day of working. We published the first image in an article in *JAC,* "Queerness: An Impossible Subject for Composition," and the second in a book chapter entitled "Queerness, Multimodality, and the Possibilities of Re/Orientation." While acknowledging some good-natured (we hope!) narcissism in generating and disseminating such images of ourselves, we intend them to act as rhetorical *détournement*—on ourselves, on acts of authorship, and on authoring collaboratively. Specifically, after collaborating for many years, and collaborating predominantly about topics in queer theory and the possible implications of queer theory for the teaching of writing, we began to reflect on how we have both, individually and collectively, come to be identified in composition studies as "queer compositionists"—an identification that, once made, we have used in the production of our own scholarly work that, once made, we have been called upon by the profession to reproduce.

The first image, Jonathan in bed with himself, parodies Jonathan as a figure in composition studies known for writing about his own identity, his own queerness, as possibly relevant to the teaching of writing. While we both understand the necessity of talking about sexuality vis-à-vis writing and its teaching, particularly given the tight braid that the social turn has woven around identity and composition, we cannot escape a sense of making spectacles of ourselves in writing so frequently about the relevance of our queerness to our work. To keep that relevance enlivened, to keep it from becoming a stale reminder of yet another identity that must be acknowledged

in the charm bracelet of composition's embrace of identity politics, we have felt the need to create ever more spectacular statements and images, even risking some self-parody—Jonathan loving himself so much, reproducing himself and his queerness in his works again and again. But a contradiction arises immediately as Jonathan is in an untenable position: once he announces that he is queer and that his queerness is relevant to his work as a compositionist, he is called upon by the profession to validate that connection, calling attention to it, arguing it in different ways, and generally representing the queer. Indeed, both of us, sometimes even together, have been called upon to comment again and again on the queer and its relation to composition. The Conference on College Composition and Communication has tasked us with different committees (in good faith, we believe) to speak about queerness within the profession, calling on us repeatedly to represent "our" interests.

Such calling seems, at times, inescapable. At a Computers and Writing conference a few years ago, when we were giving a presentation about trauma and writing studies and never once mentioned sexuality or queerness, both of us were nonetheless asked by a major scholar in the field to explain what our presentation had to do with queer theory. Nothing, we replied; it had nothing to do with queer theory. But we clearly heard the call to articulate the queer, to offer, yet again, a first-person iteration. That's how much our very presence, our embodiedness as speaking subjects and publishing authors in the field, has become associated with queerness. The ceaseless call to validate such connections seems itself excessive, a point parodied in the picture. As such, we pair the spectacularity of the image in the *JAC* article with the phrase *Ceci n'est pas un auteur composé,* a riff on Magritte's picture of a pipe announcing that "This is not a pipe." We give the photo manipulation of Jonathan yet an additional queer twist by calling attention to the excessive and parodic queerness of the photo, thus questioning the stability of Jonathan's own identity as a "queer compositionist" and his relation to the project of writing about queers writing about writing and writing instruction. This "queer compositionist" is a representation, an identification within the discipline, that we both embrace and resist, one to which we have—and attempt to articulate visually—a complex, contradictory relationship.

Such complexity and contradiction inform our relationship to collaboration as well, and we use the rhetorical affordances of photo manipulation to articulate that relationship in the depiction of ourselves in bed together. The image riffs in part on a subject of much of our writing—queerness and sexuality—but it also visually performs a bit of queerness: a gay man and a lesbian are in bed together, clearly relaxed and intimate, with Jackie, a woman, possessing some startlingly masculine characteristics, including chest hair and the "power" of holding the remote. Maleness and femaleness, typical markers of gender role and power differentials, get mixed and remixed in this manipulation to show us both embodying and exercising power (through media technologies, no less!) while also

clearly comfortable and intimate with each other. We use the visual here to depict our collaboration as disrupting the heteronormative spectacularity of traditionally gendered power relations. We used this image for our chapter in Arola and Wysocki's *Composing (Media) = Composing (Embodiment)* as an example of how the affordances of multimodality allow us to articulate the many reorientations that collaboration prompts.

More to the point, though, we find collaboration itself an act of excess, and necessarily so. Any authorial collaboration tries to find *a* voice among the many varied interests and possibilities that its collaborators bring together. But in our experience, thinking and writing collaboratively has typically generated, for any one project, much more text than we can use. This excess text sometimes finds itself in digital trash cans, and sometimes generating other projects, collectively and individually. Our collaboration has come to depend on the availability of that excess, on the meeting of minds with numerous thoughts and desires that become entangled with one another in the production of thoughts and texts that could not otherwise exist individually. But collaboration is also necessarily queer in its excesses, in that successful collaboration breeds more collaboration with more parties—not just the tightening of an initial collaborative relationship (though often that too). That is, good collaborators are often promiscuous, even as they enjoy their intimacy with one another. In learning to collaborate with each other, we have also learned to collaborate with others, and we eagerly embrace the pleasures of multiple relationships, stepping out on each other to play with others while always knowing we have a home with the other. We metaphorize the *techne* of collaboration through sex because, like having sex, writing together involves both skill and art, the development of working relations but also a sense of how to move with each other, when to stroke with encouragement, when to slap, gently, with critique, but inevitably a critique that soothes into the production of better text. Our images of collaboration embody the very embodied nature of working together. Moreover, we might not have been able to reflect critically on our own collaboration without having first designed the images, creating *détournement* on our figuration and self-perception within the profession.

While not everyone in the field will jibe with our sexual metaphors, we believe they offer a glimpse into why many of us value collaboration (it feels oh so good) but often shy away from encouraging it among our students (lest they become too promiscuous). Many of us worry incessantly over the promiscuousness of students and their texts, with plagiarism marking the very worst kinds of textual relations students can have. However, *we* insist on a certain promiscuity because the promiscuous, as both metaphor and a kind of excess, gestures toward the illicit, the approach to the taboo. Yes, students need awareness of intellectual property and appropriate citation. But thinking and designing promiscuously shouldn't be forgotten in the rush to set appropriate boundaries around relationships among "texts." The promiscuous challenges us to think in and through multiplicities—of text, of different kinds of "texts," of the identities of those who produce and read "texts," and the many disparate desires that we have for the production and consumption of texts.

DESIGNING EXCESS: THE EXCESSIVE CLASSROOM

Photo manipulation is just one of the capabilities of textual promiscuity, of what others have called remixing, the bringing together of different "texts" to generate new insights. But visual rhetoric in composition scholarship remains largely interpretive, teaching us how to teach our students to analyze—and even compose—visual texts without necessarily muddying our own hands by creating texts. Certainly, there are important exceptions: Wysocki, Vielstimmig, the work of various coverweb editors of *Kairos,* etc. And just as certainly, there are good reasons why we seem to have stalled ourselves as prosumerist teachers. Given the amount of other work that must be done in the writing classroom (and in our professional lives), how do we carve out time for our students' critical production of multimodal texts? How do we find time for our own? Who will buy the software? How will we gain knowledge for ourselves not just of the technical know-how to run Photoshop but also of the unique logics and rhetorical affordances therein? Most important, how do we *do* multimodality on its own terms rather than as a supplement to our own essayistic predilections? We can start by teaching the interpretation of images more robustly, with more attention to productive excesses of image and media manipulation. An example from the master of media manipulation, David Lynch, is instructive.

Early in Lynch's film *Mulholland Drive,* one man tells another that he's been having nightmares about a beast behind the dumpster at Winky's. His friend takes him outside to the "real" garbage can, ostensibly to show him there's nothing there, but to his surprise (and our own), the horrifying face that has haunted him is really there. The man dies of a heart attack. In this case, several narratives are disrupted: (1) the logical, in which we expect more-or-less delineated genres (no elements of fantasy will intrude on a "realist" scene); (2) the anticipatory/false horrific, a convention of horror movies, in which we expect a "decoy" horror (fearing a monster, the heroine finds her cat knocking over the garbage can); and (3) the therapeutic, in which we expect that any phantasm can be removed by attempting to "see" it (or its absence) in our waking life.

Combined in the waking-dream-within-a-waking-dream, these disruptions of narrative desire open wide the incommensurable fear and arousal that attend the "sudden unknown." The Winky's scene is a slap-in-the-face spectacular reminder of our own never-sated desire—it enacts our horror at discovering that something we thought of as "known" is not known, has never been known, cannot be known. We'd like to suggest that the scene from *Mulholland Drive* echoes our reading of the spectacular queer body and the ongoing desire to know—to own—that body. When we encounter the monster face of the familiar—a face that we have ourselves constructed—we seek to "speak the fear," our last charm (*mano fico*) against the evil eye. But because we cannot own (but can only desire more), we discover our charm is laughable. Just as the monster in the man's head is really there, the horror of the abject queer body is really there. The visceral horror of that scene comes from our even deeper fear that our constructions might be "real," that they might take on body and flesh—and that they will enact a terror that we ourselves have constructed. They might be as horrible as we have imagined them to be! Further, the terror of the Other's body produces such an excess of feeling in us that we are made aware, too, of our own body. As Monika I. Hogan has written in her discussion of ethical contact, this excess of feeling produces an awareness that challenges the denial of the body that holds our fear in place; "ethical contact" means feeling not just *moved* but *implicated*—"not only a witnessing consciousness, but a connected and vulnerable body as well." The spectacular queer is the manifestation of that vulnerability; the ridiculed *mano fico*; our own monster behind the Winky's dumpster. It is this same multiplicitous disruption that stirs the uneasy reception of this queer *techne*: this *techne* asserts that, as spectacle, the queer body and representations thereof cannot be "owned" or known.

We certainly understand some reluctance on the part of our field to fully embrace postliterate multimodality. At the same time, we believe, with Kinney, Girshin, and Bowlin, that a "third turn" of the social is here, and critical production of new media texts forms a key part of that turn. We further believe, with Sirc in his "Box-Logic," that composition as a field has privileged the "composed" text over less-reasoned but still-powerful sorts of texts. Some of those texts must be the messy, dis-composed, excessive texts—the "texts" that play with the spectacularity of representation and circulations of desire—that new and multimedia make possible.

How do we move forward, if not through?

How do we make sense and feeling of a radically expansive world of images given the DIY incursion into public authoring? As more of us engage our students in constructing new media texts, writing blogs, or just participating in Twitter feeds or Blackboard discussions, we are discovering the heady intersections of text and identity that we knew were there but had not been able to (quite) (always) make visible. These networked technologies make it easier to do such things; we resist, however, the idea that the technologies are all that have made it possible. As Wysocki writes, "Computer technologies heighten our awareness of the visuality of texts" (59). In any number of our courses—graduate and undergraduate—because of inspiration, exploration, and sometimes natural disaster (flooding in the computer classroom), we use representational technologies that are nonnetworked/low-end (colored markers, collage) and nonnetworked/high-end (Photoshop). At other times, the network is very much our aesthetic friend; the point is to facilitate access to the means of representation and distribution. It is emphasizing, simultaneously, critique and design.

Again, with Sirc, we want our students (and our fellow teachers) to take an "art stance to the everyday" ("Box-Logic" 119), and like Sirc, who embeds his discussion of new media within an exploration of Walter Benjamin's *Arcades Project*, we want to encourage a critical understanding of that stance. Our own stance led to our "queer twist" on Situationist aesthetics and an expanded sense of rhetoric and *technai*. What might others' stances lead to? What might others find at the moment of instantiation, when (teacherly or student) self encounters technology? For us, that moment of instantiation—of the flesh made real—seems ripe for rhetorical and embodied action. It also encourages a sense of material connection to text, and a deployment of the aesthetic through the material, as writers, artists, and designers offer their work as a physical interruption of alienated representation. Finally, it encourages us to put our bodies on the line, to "risk" spectacle. Linder writes of Situationist cultural studies that "yes, we can act in the world, and let me speak of bodies, pleasures, and paradises lost to suggest that it happens all the time. The real trick, it seems, is not in reaching Paradise, but staying there" (370).

4

Collaboration, Interactivity, and the *Dérive* in Computer Gaming

IN THE PREVIOUS TWO CHAPTERS, WE HAVE considered how the multimodality of working with video and photo manipulation opens up different rhetorical affordances that exceed the textual forms of communication that we as compositionists have spent much energy studying and supporting. The aural, visual, and affective domains of multimodality demand our attention. But multimodality can be understood not just in terms of the materials that people use to compose, but also in terms of the increasingly collaborative and interactive ways in which composers engage media spaces as sites of literacy development, play, and experimentation. In *Multimodality*, Gunter Kress argues that "all communication is movement. Interacting in dialogue is movement: my interest directs my attention; it frames a part of the semiotic world. I receive and construct a complex sign/text/ensemble as my response in return" (168).

We might catch a sense of Kress's argument in the following brief summary of an instance of computer gameplay: *A warlock prepares a spell that will deftly take the enemy knight by surprise. As he clicks icons to ready his potion, a fellow player types that she's just been ambushed by a rogue with a powerful invisibility cloak. To protect her guildmates, K'thora warns Gar'n and others in the area of the imminent threat. Multi-colored text flashes across screens as players exchange information, manipulate icons to cast protective spells, and strategize about how best to handle the situation. Fingers fly across keyboards and voices rise with anxious enthusiasm as players think quickly, assessing their options. Coordination and collaboration become key in fending off the rogue, earning points to gain experience and "level" characters, increasing the reputation of the guild, and having a good time.*

Over and beyond the multimediated nature of the computer game interface, we sense the complex moves of interaction and collaboration that are key to successful gameplay. In particular, massively multiplayer online role-playing games (MMORPGs), such as *World of Warcraft*, from which we drew the preceding scene, offer significant social networking opportunities for many students. When we consider how video and computer gaming is a major source of entertainment for many young people, as well as a popular form of socializing and community building as players interact with and get to know one another, we sense a significant arena in which to explore not just multimodal composition but also multimodal *interaction.* That is, we should consider multimodality as both multiple modes of communication and multiple paths and possibilities of communicative interaction.

Sure, some writing instructors bemoan students' involvement with such games, which sometimes divert student attention from more "academic" tasks and literacy practices, and some evidence suggests that some gamers are addicted to the games they play, or to playing, period.[1] However, many such games are textually and visually rich and require quite a bit of reading, writing, and multimodal thinking. At the most basic level, gaming often requires the complex use of multiple literacies and a need to develop a sense of how text and visuals interact. Focusing specifically on the relationship between literacy development and gaming, scholars such as James Paul Gee, Gail Hawisher, and Cynthia Selfe have published groundbreaking work calling for scholarly attention to how a variety of gaming platforms might be used in educational settings.[2]

Some in our field, particularly in computers and composition studies, have launched research projects building on the work of Gee, Selfe, Hawisher, and others. However, many in the larger field of composition studies are not yet aware of the possibilities for transforming the way we approach writing instruction that emerge when critically considering the potential place of computer gaming (with its complex mélange of text, image, video, and multimedia) in the writing classroom. In this chapter, we argue broadly, directing our arguments to those in our discipline who have not yet considered the possibilities afforded composition by computer and video gaming—the rhetorical affordances, as it were, of such multimodality. Specifically, we argue that incorporating gaming into composition courses may not only enliven writing instruction for many of our students, but also transform our approach to literacy through a strong consideration of collaboration and interaction in multimodal composing spaces.

When Jonathan first undertook research into gaming (and published an earlier version of this research), he focused his energies on how gaming could support primarily textual literacies; on further research and reflection, however, we consider in this chapter how gaming challenges us to think multimodally about literate and rhetorical practices. We argue that gaming offers us a rich venue to see *multiple* literacies—the visual, the technological, and the textual—at play. Since many games are also collaborative in nature, they provide an opportunity to see such literacies in evolving communal contexts. Turning our attention to gaming, then, gives us a significant opportunity to examine complex rhetorical work in action. More than this, however, paying attention to gaming and the "gaming lives" of our students invites students to speak with us about how literacy itself is changing. In such conversations, our approach to writing instruction may shift substantively from "introducing" students to varieties of literate and rhetorical practice to exploring with them the kinds of emerging literate practices that may be personally, professionally, and critically useful. To move in this direction, following in Hawisher and Selfe's footsteps, we want to pay close attention to the literacy narratives and compositional practices of gamers. We are interested in particular in how they articulate their understanding of how the multimodality of gaming affects their understanding of both (1) what literacy is and (2) how gaming contributes to their development as literate citizens. By analyzing the literacy practices of two gamers, as well as the practices of gamers outside of specific curricular contexts, we take seriously Selfe and Hawisher's call (in *Literate Lives*) to be mindful of our students' extracurricular literacies. Further, we want to push the discussion by putting gamers' insights about those literacies into conversation with scholarly approaches and insights about new media literacies.

As we will see, paying attention to gaming shows us how some gamers engage actively in developing high-level literacy strategies, such as reflectivity, transliteracy connections, collaborative writing, multicultural awareness, and critical thinking. Moreover, we note in particular how many gamers practice Debordian *dérive*, opening up new ways of conceiving mass market games by playing them *against* themselves and in ways far removed from the intentions of game designers. While games are certainly "useful" for literacy development in the ways that Gee suggests, their players often have ideas of their own about what to do with games—ideas that challenge us to rethink the possibilities for composing and media production in the writing classroom. Ultimately, paying attention to these literate practices shows us how multimodality can be understood as the creation of pathways for composers to interact with one another and to interact with games themselves as composing spaces.

BATTLE OF THE DOMAINS: COMPOSITION SCHOLARSHIP ABOUT LITERACY AND GAMING

Given the complex relationship that gaming has with issues of community, communication, and literacy, it has generated a fair amount of scholarly attention in fields such as communications and sociology, resulting in the emerging field of game studies. Such work usually focuses on understanding gaming in its sociological dimensions (e.g., the connection between gaming and violence) or on the promulgation of stereotypical forms of identity in gaming spaces (e.g., do video games promote sexist images of women—or men, for that matter?).[3] Scholars in English studies, particularly in computers and composition studies, turn their attention to gaming and its rhetorical registers. For example, Janet H. Murray's now-classic *Hamlet on the Holodeck: The Future of Narrative in Cyberspace* considers the impact of cybermedia on how we tell stories and create fiction. Some compositionists have been quick to pick up on the implications of such work for the writing classroom. In "Connecting Video Games and Storytelling to Teach Narratives in First-Year Composition," for instance, Zoevera Ann Jackson maintains that we should pay attention to how a knowledge of gaming can be used to engage and expand students' understanding of constructing and writing narratives: "By offering the use of video games as a way to give students a better understanding of storytelling concepts, I hope to give teachers an alternative way to have students write narratives in the first-year composition course."

Beyond engaging "student interest" to teach the literacies that we think are important, we should consider how complex communications such as video and computer games might actually *shift* what passes as "literate" in our society. Seeing gaming as a compositional space allows us to appreciate the rhetorical capabilities of that space—not just the "what" of gaming, but the "how," "why," and "to what end?" Two of the most recent influential books that pick up just this issue have been Gunther Kress's *Literacy in the New Media Age* and James Paul Gee's *What Video Games Have to Teach Us about Learning and Literacy*. Kress and Gee are useful here because they are two of the leading scholars who have focused attention on multimodal texts, and Gee in particular has claimed provocatively that gaming has much to teach us about learning and literacy. Kress maintains throughout his book that the "dominance" of the image and the screen among younger people will reshape what passes as literacy. This moment is not the *end* of literacy, though. Kress acknowledges that many, if not in some ways all, of the technologically enabled communications platforms prompt at least some kind of writing—and sometimes a lot of it. For Kress, the primary issue facing literacy scholars is the situation of writing vis-à-vis other modes, such as the visual, and he worries about the "question of the future of writing": "Image has coexisted with writing, as of course has speech. In the era of the dominance of writing, when the logic of writing organized the page, image appeared on the page subject to the logic of writing" (7). At this historical juncture, however, writing's former dominance is being called into question. Throughout his book, Kress unpacks some of the features of this shift, but he refrains, unlike Sven Birkerts or Neil Postman, from passing judgment on and waxing nostalgic about the move from the dominance of writing to other modes. For Kress, the end result of this change in writing's dominance is clear: "One [engagement with text] was the move towards contemplation; the other is a move towards outward action" (59–60). Kress's argument about such change has a direct relationship to our work as literacy specialists and writing instructors. How can we simultaneously pay attention to these new modes of literacy in our classrooms and also value "older" modes that we know to be useful and productive of critical thinking, of the kinds of careful and imaginative reflection that Kress, among others, associates with reading long printed texts and writing essays? Or is our work as literacy educators changing fundamentally, moving from working with textual projects to multimedia?

Similarly, Gee argues that "when people learn to play video games, they are learning a new *literacy*" (13). Like Kress, he senses significant changes in what will count as literacy in the future, and he examines specifically the impact of video and computer gaming on literacy. Gee traces no less than thirty-six different "learning principles" that video and computer games seem to promote among those playing them. These include the "text principle," the "intertextual principle," and the "multimodal principle," through which learners—that is, gamers—learn how to read, understand, and manipulate a variety of texts in a variety of circumstances. For instance, to play effectively, gamers have to learn how to "read" images and text (in a chat box, for example) both independently and in relation to each other. In this way, learning and literacy become multimodal, just as Kress suggests. However, Gee also writes that participating in gaming can promote *critical* learning. Specifically, he argues that "critical learning . . . involves learning to think of semiotic domains as design spaces that manipulate us . . . in certain ways and that we can manipulate in certain ways" (43). He argues that gamers will learn all the more effectively and powerfully since they master the skills necessary not only to game but also to experiment with the rules of the games they play, creating new skills and literacies in the process.

While Kress and Gee have been among the most influential voices in turning our attention to the impact of new media technologies on our students' and our own conception of literacy, their insights are grounded for the most part in their own observations. For example, what is often missing in Gee's discussion in particular is a closer "paying attention" to what students and gamers *themselves* perceive as significant learning and literacy experiences and developments as they game. Gee envisions good games as embodying good principles of learning; but what do students make of their engagement with such games?

To a great extent, Selfe and Hawisher's *Gaming Lives in the Twenty-First Century: Literate Connections* (2007) answers this question. Their approach is unique in the fields of computer and composition studies and gaming studies in that they asked contributors to their volume to collect "life histories" of gamers, who reflect on what gaming means to them and their literacy development. Specifically, the essays in the collection "explore the differing perspectives [on gaming] by relating the stories and experiences of individual gamers who have formed their own observations about the benefits and shortcomings of game playing in a digital world" (3). Chapters focus variously on the "social dimensions of gaming" such as relationship and community building during gameplay; "gaming and difference" as revealed through the experiences of gamers coming from a variety of racial, gender, sexual, and age backgrounds; and intersections between "gaming and literacy." Selfe and Hawisher also asked each contributor to make connections between the life histories they collected and potential ramifications for the teaching of writing; their hope was to "provide literacy instructors with a methodology that they might use in their own teaching for evaluating the impact of gaming on literate lives" (3).

This work builds on that of Kress and Gee, and Selfe and Hawisher make it clear in their introduction that they wish to extend Gee's work in particular, exploring his claims about the efficacy of gaming in helping students develop certain kinds of literacy and critical thinking abilities. In their article, "Computer Gaming as Literacy," Cynthia Selfe, Anne F. Mareck, and Josh Gardiner conclude that "young people's literacy activity in the semiotic domain of gaming may prepare them to operate, communicate, and exchange information effectively in a world that is increasingly digital and transnational—and in ways that their formal school does not" (30). Such statements corroborate Gee's assertions. The methodological gain in *Gaming Lives* is that such claims are grounded in analysis of the experiences of actual gamers. A potential drawback is that connections between the gamers' experiences and the writing classroom are often left underexplored.

To fill in this gap, many teacher-scholars have reflected on the process of incorporating computer and video games into the writing classroom, and they have created an ingenious set of pedagogical possibilities. For instance, Max Lieberman, writing for a 2010 special issue of *Currents in Electronic Literacy*, summarizes the relevant literature on the subject and discusses four ways to teach with video games. Two simple approaches involve, on the one hand, teaching games with particular content useful to the writing classroom and, on the other hand, using games themselves as the "text" of a particular class, usually a thematically driven comp course. Lieberman argues that "the most straightforward way to teach with video games is to have students play a game containing content that aligns with an existing school curriculum." Thinking about games as complex, multimodal texts in which we deploy a variety of interpretive strategies can provide students with engaging material on which to practice critical reading and writing. Along such lines, Edmond Y. Chang, writing about MMORPGs in "Gaming as Writing, or, *World of Warcraft* as World of Wordcraft," picks up on Selfe's call to broaden the communicative domains of the writing classroom by thinking of complex games, such as *WoW*, as complex "texts." MMORPGs offers students rich opportunities to think critically about what defines "game" and "play" as not only creative but also critical engagements. Students can also examine narratives of games, as well as develop what Chang calls "critical playing":

> Students can close read and analyze how cultural formations and ideological assumptions like race, gender, class, sexuality, citizen, and nation are rendered and enacted by a game;

> students can analyze how formations, stereotypes, and logics of race, gender, class, and so on are generated by a game not only through visuals but through text, chat, actions, and so on; students can analyze their avatar's "backstory" for cultural assumptions and stereotypes; students can analyze the discourses about video games, particularly the controversies over the dangers of video games or the pedagogical value of video games; students can close play and analyze particular aspects of a game like character creation, quests or rewards, mise en scène, even the packaging of and written materials for the game; students can close play and critique "serious games" from sites like the Serious Games Initiative (www.seriousgames.org/) or Water Cooler Games (www.watercoolergames.org/); students can close play and critique overtly and problematically political or ideological games like the US Army's game America's Army (www.goarmy.com/aarmy/index.jsp); students can play games to look for exploits and ways to resist the social or ideological logics of the game; students can look at critical responses to video games in scholarship, media, and art, particularly at groups like the Radical Software Group (r-s-g.org/) or Velvet Strike (www.opensorcery.net/velvet-strike/) or Critical Art Ensemble (www.critical-art.net/); students can propose or create or write their own "serious" video game.

Indeed, the possibilities for thinking of games as rich texts seem endless.

Lieberman pushes beyond such text-driven approaches, however, in that he argues for the possibility of constructing classes using what he calls "game-like motivational systems," in which the course itself is structured as a kind of video game, with rewards offered for completing tasks. Even more provocatively, Lieberman pushes the bounds of the traditional composition class when he suggests that students could actually create their own games:

> Making a video game can be an expensive, time-consuming process requiring skilled programmers, artists, game designers

> and writers. Yet research shows that with the right tools and technical assistance, students are capable of creating video games as part of their education, and that they enjoy and benefit from the experience.

Lieberman argues cogently that designing a game, particularly an interactive game, can offer students significant development in both technical facility and thinking creatively and critically. Further, while he admits that "the implementation of this approach requires a greater willingness to restructure an existing curriculum," he maintains that "theories of constructivist and constructionist learning . . . hold that learning is most effective when students construct mental models or tangible artifacts."

We are tantalized by Lieberman's call to think of the authoring of games as a useful compositional process, even as we are left wondering what such a pedagogy might look like. Part of our interest lies in embracing in earnest the call to think (and compose) multimodally that is embedded in Lieberman's provocation. It just as surely lies in our thinking that games have probably been treated too "textually" (as in Chang's case), with less attention paid to their multimodal affordances. Jonathan Alexander and Elizabeth Losh, writing in 2010 about the use of computer games in writing courses, note that

> since the publication of Gee's book and the popularization of its principles by Henry Jenkins and many others, a number of researchers—many within game studies—have begun to question how the literacy movement has attempted to colonize game culture by channeling subversive behaviors into supposedly productive and normative conduct.

Along such lines, we note the tendency in many comp scholars' work on gaming in the writing classroom to fall back on textual practices as the seemingly necessary way to engage video and computer games. For instance, Christopher Paul, writing in "World of Rhetcraft: Rhetorical Production and Raiding in *World of Warcraft*," notes the multimediated nature of gaming while ultimately emphasizing games as essentially textual:

> Online games are just one more of the many ways in which social interaction is becoming increasingly digitally mediated. MMOGs stand as excellent examples of the dependence of such games on rhetorical production and consumption. No longer simply limited to what "comes out of the box," these games are regularly updated, persistent worlds inhabited by millions of people who are encouraged to interact with each other. Raiding is one way in which players have chosen to play MMOGs and raiding, as a practice, is fundamentally rooted in reading and writing. (159)

We cannot help but ask, though: what lies beyond, or at least in addition to, reading and writing in working with games?

Alexander and Losh note that at least two dominant schools of thought have arisen in game studies about the pedagogical usefulness of games:

> As classroom or lab-based projects aimed at developing game literacies in K–12 education are launched, a schism in pedagogical philosophies has emerged. One camp, largely associated with Gee and the University of Wisconsin (Squire, Games, etc.) and the University of Arizona (Hayes, Robison, etc.), largely focuses on teaching game design with an emphasis on creating interfaces and user experiences with simple authoring tools, while another camp, largely associated with what Michael Mateas and others have called "the Georgia Tech approach," claims that "procedural literacy" requires an ability to read lines of code in specific programming languages.

At some point, as Alexander and Losh's schematic overview suggests, teaching the composing of games bleeds into teaching coding. We want to pause for a moment before committing ourselves to such a course. After all, we recognize that very few composition instructors will have the ability, or desire, to teach code. So, if we ask that we as a discipline stretch ourselves, even stretch significantly, we recognize a breaking point; and we assert that, however important knowing how to code is, the teaching of coding is justifiably the domain of another discipline—and we can partner with that discipline.

At the same time, teacher-scholars in composition studies note the need to "up the ante," as it were, on our technological expertise. Rebekah Shultz Colby and Richard Colby discuss in "A Pedagogy of Play: Integrating Computer Games into the Writing Classroom" the need to engage students' interests when having them write about gaming (such as allowing economics majors to write strategy guides about the imaginary economies of MMORPGs); however, they also note the need to allow students to pursue a variety of technologically rich platforms and modes of communication in thinking, learning, and composing about games. For instance, they discuss the case of "Tiffany, who enrolled in the course because it was convenient and her roommate was taking it, [but who knew] very little about the game. Both she and her roommate Liz, an experienced *WoW* player, often participate in the social networking website Twitter. They [decided] to create a proposal for a Twitter-like website that tracks what certain players are up to in a game guild" (309). Along similar lines, they note that "students might also decide that they want to create a class-wide guild, all starting from level one, seeing how far they can get within the term. They might create guild websites, strategy guides for lower-level content, and even systems of organization for making sure classmates get to run a particular instance or complete a quest. Thus, students approach the puzzles of the gamespace by attempting to solve them rhetorically" (309). Abby M. Dubisar and Jason Palmeri are perhaps even more insistent in their article, "Palin/Pathos/Peter Griffin: Political Video Remix and Composition Pedagogy," about the need to open ourselves as comp instructors to new technologically driven communication platforms:

> We should craft flexible assignments and activities that account for the diverse activist and technological literacies students bring to class, taking care to avoid assumptions about what students already know. As Susan's case shows, for example, students who are apprehensive about technology may be able to gain confidence by being able to draw on other literary practices (such as historical research) to craft projects that they find meaningful. Furthermore, Susan's case reminds

> us of the need to make explicit technology instruction a part of class—conducting workshops in which students can view a demonstration of how to use necessary software and then receive help from peers and instructor as they experiment with the software themselves. (88)

Supporting students in such endeavors is certainly a challenge—but not just a technological one. Beyond teaching software, we need to be cognizant of the particular rhetorical capabilities of multimodal texts and what different composers might be able to do with, and in the case of games, *within* multimodal spaces. Games, especially complex, highly interactive games such as MMORPGs, pose a particular challenge for compositionists. As we have seen, some writing instructors are attracted to gaming as subject matter. What about games as compositional spaces themselves, though? How have games become spaces in and around which students compose?

For the remainder of the chapter, we use Selfe and Hawisher's approach and extend it by showing how attentiveness to gamers' experiences and reflections might directly impact—even alter—specific composition pedagogies. We discuss the content of interviews Jonathan conducted with two students who actively play a variety of computer and video games and use numerous other new media communications platforms, such as email, texting, and the Web, to aid and augment their play. While recounting their comments, we examine their thinking and reflections about their own literacy practices in light of what thinkers such as Kress and Gee suggest about such practices. In the process, we see that these students understand their literacy practices in ways that are both comparable to and divergent from those suggested by both Kress and Gee. For instance, we want to modify Kress's metaphors of "dominance" and his assertion that we are now experiencing "the broad move from the now centuries-long dominance of writing to the new dominance of the image" and likewise "the move from the dominance of the medium of the book to the dominance of the medium of the screen" (*Literacy* 1). While we believe there are definite *shifts* taking place, we remain unconvinced that image and screen are becoming "dominant," even though we may have to adjust our thinking about their interaction, as well as how complex gamespaces often require a high degree of collaborative authoring and meaning negotiation. Second, we examine the work of gamers outside the comp classroom to see what kinds of rhetorical practice they engage in and how they understand the games such as *WoW* as spaces for critical thought, reflection, and action. Specifically, we are interested in how the gamespace itself becomes for some gamers a robust compositional space for experimenting with different rhetorical positions and possibilities. We conclude with a discussion of how our composition pedagogies may shift as we pay attention to and consider more carefully students' observations and reflections on their own multimodal literacy practices.

BECOMING REFLECTIVE ABOUT LITERACY, TRANSLITERACIES, AND COLLABORATIVE WRITING

First, let's consider the amount and kind of composing occurring in and around gaming.

Two undergraduate students at the University of Cincinnati agreed to share with Jonathan some of their insights about gaming and even introduced him to several of the games they play, including one MMORPG, *World of Warcraft*. The students, young (twenty-one years old) white men named Mike and Matt (pseudonyms) were music majors who claimed to be lifelong gamers.[4] Both spoke consistently of the importance of being "immersed" in a gaming environment, particularly if the game was an MMORPG; such immersion relies on the use of multiple types of technologically enabled forms of communication, including cell phones, instant messaging (IM), and programs such as Ventrilo or TeamSpeak, with which these gamers communicate as they play. Shortly after the initial interviews, both students contacted Jonathan about using Skype, which allows computer users to chat telephonically with one another through the Internet, to complement and enhance their game play. Their technological sophistication—and *technological literacy*—is remarkable. Neither student studied computer science, but both are fairly representative of members of their socioeconomic class who are also interested in gaming: they use multiple communication technologies during play, and they do so with no small amount of skill. During gameplay, they simultaneously manipulate icons, chat or text via cell phone or Skype, and type in chat boxes to coordinate play and complete quests. We received a taste of the high level of multitasking of which these students are capable when Matt came to Jonathan's graduate seminar on electronic literacies to give a demonstration of *World of Warcraft*. He brought his wireless laptop and positioned its screen on a document camera so the screen could be projected up for all to see. He then simultaneously played the game, chatted with other players who noticed he was "on," and talked his way through what he was doing. It was an impressive sight. In fact, Mike suggested to Jonathan, as the latter learned to play *World of Warcraft*, that Jonathan focus his attention on the text chat boxes. He maintained that almost everything Jonathan needed to know and pay attention to as he was playing was occurring in the chat boxes, of which there were as many as three at one time—one for individual communication, one for guild communication, and one for communication with players he might be playing with directly in groups to complete quests—all text being color coordinated as well. Mike went so far as to suggest that knowing how to move quickly between the chat boxes and the icons, needed for spell manipulation, was crucial in successful gameplay.

Intense communication is not confined to the gamespace, however. Mike, the leader of his guild in *World of Warcraft*, worked with his long-distance friend Josh to set up a website to organize guild events, coordinate play, and share information with other guild members. Communication on the website is mediated largely through numerous message boards, and posting weekly on the boards is required for continued membership in the guild. Most players in the guild post regularly and enthusiastically. Member participation and discourse on these boards offer a concrete example of what Gee calls "design grammars" in such semiotic domains—or, more specifically, "the principles and patterns in terms of which one can recognize what is and what is not an acceptable or typical social practice and identity in regard to the affinity group associated with a semiotic domain" (30). Guild members refer in specific and specialized language to various aspects of the game, and it becomes quickly apparent who is a newbie and who is not.

Along these lines, the website *World of Warcraft* Pro (wow-pro .com/) offers extensive space for posters not only to participate on message boards but also to compose (often lengthy) guides to playing the game. Other readers comment on these guides, prompting revision and development of further guides. One in particular caught our attention: Groktal's "Guide Writing" (http://wow-pro .com/node/635). In his guide, Groktal advises other guide writers about how to organize information accurately and effectively, creating introductions, including information in the body of the guide, and rounding out with a good conclusion. Along these lines, another poster, Toogie, created "A Newcomer's Guide to Warcraft—Toogie Learned the Hard Way So You Don't Have To!" (wow-pro .com/node/1077). The introduction lays out in a chatty way the author's motivation for creating the guide and its purpose:

> Hi. I decided to make this guide for fun and to kill some down time at work. I will cover many topics from character creation to strategies for making money. It's not a guide to tell you what to do but it will explain many things which should be considered when rolling your first character and many lessons I learned the hard way. I have played this game way too much so I hope this helps some people out there.

With tongue in cheek, Toogie outlines what his guide will do and makes a stab at creating ethos, a sense of why he is qualified to author the guide: "I learned the hard way" and "I have played this game way too much." Further examples of such writing are easily accessible through quick searches.

Such gamers use multiple technologically enabled communication platforms to enhance their gameplay, and we believe that Mike's and Matt's experiences also show us important literacy practices that both confirm and challenge Gee's and Kress's views of the impact of new media on literacy. Students do deploy Gee's "multiple routes" and "intertextual" principles, in which gamers understand that multiple texts and genres of texts must be used and manipulated to achieve their ends. Such students and gamers seem to be developing a degree of literacy reflectivity, in which they think critically about the kinds of communication strategies that are most effective, as well as deploy with great thoughtfulness numerous kinds of communication (from chat boxes to message boards) to organize and enhance their gameplay.

We also believe that Kress accurately points out the complexity of reading strategies involved in playing MMORPGs such as *World of Warcraft*, and he actually uses gaming as an example in his discussion of reading new media:

> The strategies for successful reading are every bit as complex as those of the written page—one might be tempted to say, more complex, given the pre-established reading path of the page—but in any case, and certainly, different. It is not that there isn't a reading path, though many games of the "role-play" variety (say, a game such as *Final Fantasy*), or even action adventure games such as the famous Lara Croft, offer alternative reading paths, something not encountered on traditional pages. Readers of such screens are used to a different strategy. (*Literacy* 161)

Given the multiple modes of textual and iconic manipulation required to "read" the game, we agree with Kress that there is certainly a significant reading complexity involved in successful gameplay. What is missing from his discussion is a more specific examination of the kinds of complexities involved. Mike and Matt provide some useful insights here.

First, both students' comments reflected what they feel playing these games offers them in terms of life skills, particularly literacies they will be able to take with them into other venues—what we call "transliteracies," or portable literacies. Matt's comments are insightful in this regard: "Mike's learning a lot about leadership [by leading the guild]. You also learn a lot about problem solving, and especially in role-playing games you have to solve a lot of problems. It can be anything from puzzles to goals." Matt suggests that effective communication, via a variety of venues and platforms,

forms an important part of critical problem solving. Interestingly, the guys see this skill as significantly tied to leadership skills that they might use in the business world or as employees in future professions—a connection we call "transliterate awareness" in that the students become more conscious of how certain literacy and rhetorical strategies might transfer to other writing environments. For instance, both Mike and Matt suggest that multitasking and engaging multimodal forms of communication are essential components of working with the new media of MMORPGs, and they attempt to link these strategies to the working world. It may be the case, given their parents' disdain for the games, that Mike and Matt overstate their case for the usefulness of these games. In many ways, though, we don't think so. Kress suggests that such multitasking and multimoding (to coin a phrase), are

> the skills of the multimodal world of communication. They entail differentiated attention to information that comes via different modes, an assessment constantly of what is foregrounded now, assessment about where the communicational load is falling, and where to attend to now. . . . [This] is reading for specific purposes, for the information that I need now at this moment. (*Literacy* 174)

Again, while we do not agree entirely with Kress's seeming insistence that reading and writing in such new media venues is completely purpose-driven, we see a lot of imaginative play in Mike's and Matt's manipulation of *World of Warcraft*, among other games, and we see them attempting to make connections about the skills they are developing and their potential usefulness outside the context of the games.[5] For instance, what we find most compelling about Mike's and Matt's reflections—and this point is one that Kress and Gee largely gloss over—is the collaborative nature of most of the writing in gaming spaces. Although Gee suggests that students often have to work together in many gaming environments, some of these students do not just "work" together but write together about a variety of topics, some game-related, some not. When Mike talks about developing a gaming strategy through email with Josh, or when both Mike and Matt refer to the activity on the discussion boards of their gaming website, they show us how writing in these contexts is highly collaborative. As Mike explains, the boards serve not only "strategic" gaming purposes, but also contribute to community and relationship building:

> The website's just like a giant forum. . . . On the website people communicate with each other about making trades and meeting times. . . . We have an announcement section where we put important announcements where only our leadership can post. . . . It's just [like] any kind of communication you would have with a friend about a book that you're reading. It's the same here, it's just like a guideline to take people, you know, to where they feel like discussing the game.

The website's discussion boards cover a variety of topics, from gaming to nongaming subjects, allowing players to communicate about a variety of topics.

We can see collaborative writing at work on a number of similar guild sites. For instance, the guild site for Thrall's Chosen (www.guildportal.com/Guild.aspx?GuildID=24923&TabID=224934) offers a "Pub" section in which players can share stories about their characters, consider feedback from other posters, and elaborate on their writing. The multipart "Bio of Kkir" has drawn much commentary and advice, suggesting story lines and improvements:

> Kkir, now an officer of Thralls Chosen, was heavily involved with conflicts all around Azeroth. Many fronts were being fought simultaneously by the guild. The battles were intense and exhausting, but largely successful. The guild had grown three fold in size and was standing up to the challenges laid out by Thrall himself.
>
> Staying busy was a good thing for Kkir. His soul mate, Kerrybella, had left the melee months ago and headed back to her tribal home. She was called home to take her place on the Heaven Light Counsel, as required by clan customs. She insisted Kkir stay and continue the important work of the guild. Kkir and Kerrybella stayed in contact via messages and

> a soul-bond few will ever understand, but the emptiness in Kkir's heart and essence weighed heavily on his life. He dealt with this by immersing himself in his work. ("Bio")

We like this brief excerpt for its direct mention of "messages" as a way for these characters (i.e., players) to stay in contact with one another.

The discussion boards on Mike's guild site also reveal how multiple individuals contribute to the development of a strategic text or "game plan" that all participants will then follow. That text takes shape through much discussion, negotiation, collaboration, and some amount of contention. Ideas are discussed, arguments put forward, rebuttals heard, and evidence (based largely on the experience of past gameplay) considered. In one instance, a board was set up to discuss how to attack a particular monster, who could only be "killed" by a collective effort of players working together. Members posted various strategies, with some writing 1,000-word "essays," introducing why this particular kill was important, setting up strategies for accomplishing the mission, considering potential counterarguments to the strategies outlined, and concluding with rhetorically rousing "calls to arms." The most intriguing aspect of such arguments, however, is that they take place as students work collaboratively on *one text*. Again, such collaboration is not uncommon in many professional fields, but we wonder to what extent we in our writing courses teach students not just to write but to write collaboratively.

CRITICAL LITERACY DEVELOPMENT AND THE GAME AS COMPOSING SPACE

Mike's and Matt's reflections on their gameplay prompt us to consider how critical these students' engagement with these games and new media spaces actually is. Such critical engagement exists both in (1) manipulating technologies to develop new media literacy fluency and (2) becoming more reflective about the complex social and cultural impacts of these new media technologies on how we communicate. Selber argues for an emphasis on critical literacy of new media, in which consumers of new media act as critical consumers; specifically, "students should . . . be able to function as more ambitious agents of positive change. In other words, students should be able to function as reflective producers of computer technologies" (*Multiliteracies* 133–34). Are Mike's and Matt's literacy practices in and on these various sites *critical* in Selber's view? At times, their reflection certainly seems so. Mike acknowledges, for instance, that gaming is "big business"; he's very aware that a lot of money changes hands as kids play these games.[6] More interesting, though, Mike discusses at one point how the game's designers have created "racial" tensions to enhance gameplay; in fact, advanced ways of playing the game revolve around racial conflict:

> [*World of Warcraft* is] a game that's focused on a war between two enemies basically, and the way the game is modeled is to make players hate people of the opposite faction and, you know, the people of the opposite faction are also players, but because they're of the opposite faction you can't help but to hate them, the way they've designed it. This is because there is no way you can talk or communicate at all to those people in the other faction. They don't understand your language; they don't know what you're saying. If you talk to them, everything you say comes through a translator and it turns it into like orcish or elvish. People of a different race, in the opposite faction, aren't going to understand anything you say. [The game designers] throw many obstructions between the two races. There's just a lot of animosity built towards each other.

Gee argues that "reading and writing should be viewed not only as mental achievements going on inside people's heads but also as social and cultural practices with economic, historical, and political implications" (8). Although Mike's reflections are mostly descriptive in the preceding passage, he articulates clearly how a lack of understanding, compounded by an inability to communicate, has the potential to enhance conflict. Certainly, the game designers understand this potential and have introduced this element into the game's war. However, Mike knows what's going on. He perceives not only how the game has been set up but also how communication is crucial in both play and in understanding the social, cultural, and political "story" generated during gameplay.

Some will argue that Mike seems to understand that he plays at "race war" when jumping online and logging in to *World of Warcraft*, and that his continued interest in the game despite this knowledge suggests a callousness about issues such as racial conflict. We doubt that Mike would support a race war in "real life," based on our conversations with him. Granted, Mike stops short of critiquing the game, which is hugely popular among gamers like himself. Still, his awareness of how the gamemakers have designed the game so that communication is figured as part of and even crucial to the game's politics seems to us pretty savvy. It is a reflective gesture—one that is willing to imagine what the designers had in mind, to refuse to be simply, in Gee's words, one of the "dupes of capitalist marketers" (205).

In an interview, Mike spoke eloquently about the importance of learning during gameplay how to communicate across multicultural differences and through multiple styles of culturally inflected communication. Players interacting with one another through MMORPGs likely encounter both nonnative speakers of English (who often play, nonetheless, on English-speaking servers) and players from diverse parts of Global North. This situation sometimes presents unique communication challenges that gamers feel are significant to work through to coordinate efforts and enhance the gaming experience:

> Something I've dealt with lately is, you know, we have actually this sixteen-year-old French Canadian kid on our guild now and he just clashes with everybody. . . . This is his secondary language, so some people when they first meet him can either be offended by some of the things he says because he hears other people say them and he'll just say them. Like, for instance, first when he got on the guild he started using a lot words like *fag* and *gay* and stuff because he saw so many of the other players using these words and stuff in negative situations calling each other this and, you know, on our guild we have a few gay members. That's not really acceptable using

> those terms in a negative connotation, so I took him aside and I explained to him, you know, I know that this isn't your language, but you can't just go saying that word like that. You're going to offend people. He said, "What does it mean?" And right when he said that I realized he wasn't, you know, trying to use any negative terms against anybody in a way that would offend someone having to do with their sexuality. [So I helped him] express himself in ways that wouldn't offend people, and teaching him what words are offensive and which aren't is really important.

Indeed, Gonzalo Frasca, in "Videogames of the Oppressed," argues that games cannot change reality but that "players can realize that there are many possible ways to deal with their personal and social reality. Hopefully, this might lead to the development of a tolerant attitude that accepts multiplicity as the rule and not the exception" (93). Given the example of Mike's awareness of *World of Warcraft*'s design, as well as his working with the French Canadian player to use more appropriate and tolerant language during gameplay, we believe that Frasca may have a strong point.

Other examples abound. For instance, The Paragonian Knights (www.guildportal.com/Guild.aspx?GuildID=30032&TabID=269178) make it clear that "hate speech" is not allowed among its members:

> We expect all members to be courteous to each other and treat others as they want to be treated. We do not tolerate bullying, flaming, discrimination, and excessive use of foul language. This includes in-game and website conduct. For the first offense we will issue a warning. Thereafter any successive transgression will result in immediate removal from the SG and Guild portal site and will not be re-invited to the SG.
>
> Please remember that you are a representative of our Super Group. Be kind and courteous to others as it will reflect back on the Super Group as a whole when you do or do not. Wear the Paragonian Knights and Rogue Knights name proudly!

Given such comments, we concur with both Gee and Kress when they maintain that such games and new media experiences can promote not only a toleration of and even interest in cultural difference, but also an understanding of the role of communication in mediating that difference and the role of literacy in working collaboratively with cultural differences in mind. Gee argues, for example, that one major learning principle gamers glean from gaming is the "Cultural Models about Semiotic Domains Principle," in which "learning is set up in such a way that learners come to think consciously and reflectively about their cultural models about a particular semiotic domain" (211)—in this case, the domain of the MMORPG. Mike is willing to analyze the semiotic situation and assist another player in understanding how text and words circulate socially and culturally in this particular gaming forum. Thus, he exhibits a fair amount of reflection about the social dimensions of language.

Do game designers consciously inculcate particular values in their game designs, or do their narratives (such as story lines involving race conflicts) arise out of a political unconscious, out of a fairly unreflective mirroring of the world around us? The answers, certainly beyond the purview of this chapter, are complex and even game-specific. We believe, however, that Mike's awareness of the game's design suggests a reflective practice that challenges Kress's seeming assumption that new media spaces promote more action than contemplation. In this particular case, these gamers engage in more than a bit of critical reflection, in a variety of ways. We see a potential intensification of such critical reflection among gamers who conceive of the gaming platform itself as a compositional space.

Further, in addition to thinking about gaming as a space to which students such as Mike and Matt *bring* a great deal of compositional practice, we should also consider how gaming platforms such as *World of Warcraft* can themselves be robust composition spaces in which to author—and authorize—different kinds of thinking and being. In *My Life as a Night Elf Priest: An Anthropological Account of* World of Warcraft, cultural anthropologist Bonnie Nardi argues that

> WoW, and other social games, are emerging as global artifacts that appear to sustain, in vastly different cultural contexts, alternatives to, or displacements of, traditional media. The narcotized populace passively immersed in a spectacle of images seems to have given way, in part at least, to activity in digital worlds that create their own playful problems to be solved with cultural imaginings such as gold raids and Dragon Kill Point systems. That people behave badly in virtual game worlds, requiring player-developed social controls, is perhaps nothing more than an indication of the worlds' status as, and footing in, real human social activity—both East and West. (196)

For an anthropologist such as Nardi, the sheer interactivity required by an MMORPG provides a rich domain through which to see—and experience—the creation of norms, the establishment of codes of conduct, the exercise of agency, and the self-policing of groups. The rich multimodality of the space facilitates complex human encounters—through text, sound, and images.

Kress understands the potential implications of analyzing and using such encounters in pedagogical situations: "Literacy and communication curricula rethought [with interactive new media in mind] offer an education in which creativity in different domains and at different levels of representation is well understood, in which both creativity and difference are seen as normal and as productive. The young who experienced that kind of curriculum might feel at ease in a world of incessant change" (*Literacy* 121). However, Gee and Kress think in terms of what the games and such new media offer students in *academic* contexts. Curiously enough, Mike and Matt see games as offering them a variety of literacies that are potentially useful outside the classroom. We believe they are beginning to understand their literacies as malleable and in need of "customizing" to meet a variety of different rhetorical situations in the world at large. In so many ways, this rhetorical awareness and sensitivity is exactly that with which many in composition studies hope to equip their students. We are unsure, however, about the extent to which we really provide our students such robust platforms for rhetorical engagement. By paying attention to such reflections, we learn about, in Gee's words, how these students "read" these games—and about the literacy practices they perceive as significant and useful.

Beyond such multicultural education, we wonder what other possibilities exist in these games for exploring alternative ways of being, for using the multimodality of the MMORPG for thinking differently. How might the rhetorical affordances embedded in *WoW,* for instance, be deployed by gamers to trace alternative paths, using the gamespace as a compositional platform in ways not necessarily intended by the game designers? Nardi's approach to *WoW* is pretty "straight," in that she focuses her attention on play within the confines of the game. But what if we think of the game as an authoring space, a hyped-up, multimedia version of MSWord?[7]

Some game theorists have begun to make such connections, most notably McKenzie Wark, whose *Gamer Theory* attempts, in however an attenuated fashion, to reenvision complex gaming platforms through the long history of avant-gardist compositional practice. On the one hand, he acknowledges the tendency to think of gaming in terms of the game: "So this is the world as it appears to the gamer: a matrix of endlessly varying games—a gamespace—all reducible to the same principles, all producing the same kind of subject who belongs to this gamespace in the same way, as a gamer to a game" (15). On the other hand, citing the Situationists, Wark asks how we might think of the game from the vantage point of play, not just the game's rules:

> What would it mean to lift one's eye from the target, to pause on the trigger, to unclench one's ever-clicking finger? Is it even possible to think outside The Cave? Perhaps with the triumph of gamespace, what the gamer as theorist needs is to reconstruct the deleted files on those who thought pure play could be a radical option, who opposed gamespace with their revolutionary playdates. The Situationists, for example. Raoul Vaneigem: "Subversion . . . is an all embracing reinsertion of things into play. It is the act whereby play grasps and reunites beings and things hitherto frozen solid in a hierarchy

> of fragments." Play, yes, but the game—no. Guy Debord: "I have scarcely begun to make you understand that I don't intend to play the game." Now *there* was a player unconcerned with an exit strategy. (15)

Ultimately, Wark sees complex gamespaces as opportunities to think beyond the game's rules, which metonymically reflect sociocultural–political "rules," and to wander, to explore, to turn, to play: "The play of meaning is made within the bounds of a game. At stake here is the relation of play to game. As topography gives way to topology, the game rises in prominence relative to play. In the realm of avant garde strategies, the game-within-constraints of George Perec trumps the play-beyond-game of Guy Debord" (181).

Embedded in such thinking, again borrowing from the Situationists, is the possibility of seeing the gamespace as a possibility for tactical play, specifically for *dérive*. We take the idea of such tactical play from *The Practice of Everyday Life* by Michel de Certeau, in which de Certeau poetically describes the

> unrecognized producers, poets of their own affairs, trailblazers in the jungles of functionalist rationality, consumers [who] produce something resembling the 'lignes d'erre" described by Deligny. They trace "indeterminate trajectories" that are apparently meaningless, since they do not cohere with the constructed, written, and prefabricated space through which they move. They are sentences that remain unpredictable within the space ordered by the organizing techniques of systems. Although they use as their material the vocabularies of established languages (those of television newspapers, the supermarket or city planning), although they remain within the framework of prescribed syntaxes (the temporal modes of schedules, paradigmatic organizations of places, etc.), these "traverses" remain heterogeneous to the systems they infiltrate and in which they sketch out the guileful ruses of different interests and desires. They circulate, come and go, overflow and drift over an imposed terrain, like the snowy waves of the

> sea slipping in among the rocks and defiles of an established order. (34)

For the Situationist International, a key tactic was the *dérive*, or "a mode of experimental behavior linked to the conditions of urban society: a technique of rapid passage through varied ambiances" ("Definitions"). More than just a stroll or a journey, the goal of such passage is, as de Certeau suggests, to make manifest "the guileful ruses of different interests and desires." In the process, norms are questioned, the possibility for expression enhanced, and alternate paths of meaning encountered and explored. What might such a *dérive* look like in gaming? Further, how might the rhetorical capabilities of gaming spaces be used to trace, in de Certeau's words, "indeterminate trajectories" that potentially question the normative and model alternative modalities of being?

Examples abound, particularly in relation to—and within—a game such as *World of Warcraft*. Indeed, exhibits devoted to player art produced in relation and reaction to the MMORPG display the variety of responses to the game; for instance, the Laguna Art Museum's "*WoW*: Emergent Media Phenomenon" allowed viewers to explore

> various forms of cultural production based on *World of Warcraft*® in particular and on gaming in general. While surveying Warcraft's fifteen-year history, the exhibition looked at artistic practices that have been influenced by game culture. The actual works by the producer of *World of Warcraft*®, Blizzard Entertainment® (headquartered in Irvine, California), provided a starting point and reference. (Laguna)

Curated by Grace Kook-Anderson in conjunction with Blizzard's curator, Tim Campbell, and artist Eddo Stern, much of the work seemed high-class "fan art," such as sculptures of *WoW* avatars, prints of different characters, and media installations of filmed individuals walking around with their names in green suspended over their heads (mimicking the positioning of avatar names in gameplay). However, some of the art embraced critique, particularly through its provocative use of the gamespace to compose *against* the rules and parameters of the game.

Perhaps one of the most substantive such enterprises is "The Third Faction: Missions from the *World of Warcraft* Art Corps," which offers an extensive blog about the Faction's in-game activities: http://thirdfaction.org/blog/. According to their site, the Faction

> is an affiliation of avatar/entities with a collective interest in exposing binary systems in Synthetic Environments. Third Faction's focus is to develop a democratized and non-hierarchical praxis as well as governance. Current members include a global cabal of artists and performers whose work has been internationally recognized and exhibited. . . . As a member of Third Faction it is your duty to participate in cross factional missions, share the missions, subvert the dominant culture, and take part in a richer experience.
>
> Although originating in World of Warcraft, the mission of the collaborative uses many methodologies to critically examine hegemonic narratives and attitudes that generally pervade game environments. ("Third")

On display at the "*WoW*: Emergent Media Phenomenon" exhibit were films and other documents that showed how the Third Faction spends time in-game trying to promote peace between the Alliance and the Horde, the warring factions of the game. Members of the Faction wander around the gamespace trying to communicate with members of the opposite race, petitioning them to cease fighting. Their *dérive* subverts the primary PvP, or "player versus player," setup of *WoW*. Inspired by Hakim Bey, the Faction has set up in *WoW* what it calls "Temporary Autonomous Zones":

> In order to elude formal structures of control an inter-species, cross-factional group of individuals banded together to form a Temporary Autonomous Zone within the World of Warcraft. Once citizens of Azeroth, this band of free thinkers fought to free a section of land in the Eastern Kingdoms from both Alliance and Horde, player characters and NPCs alike. In this new free zone they held council and voted to enact a new and temporary manner of governing the space. One that was based on a non-hierarchical system of social relationships inspired by anarchist philosophy and the writings of Hakim

> Bey. Like all TAZ, the new governance was temporary as was the space. "Any attempt at permanence that goes beyond the moment deteriorates to a structured system that inevitably stifles individual creativity. It is this chance at creativity that is real empowerment." However within the order, chaos, structure, freedom, blood and enlightenment of the TAZ, a seed was formed. An ideal that became a campaign against tyranny and war. A subset of those that took up residence in Temporary Autonomous Zone during the time of its existence formed an activist collective of artists and free thinkers, in order to campaign this campaign. This collective is known today as The Third Faction. ("Third")

The Faction uses the mechanics of the gamespace—its avatars, its interactive components, its prompting of members to form collectives ("guilds")—to create an ethos among players that is completely counter to the PvP design of the game. Or, put another way, the Faction seeks, as with the Situationists, to play the game so as to subvert its rules and, in the process, create what the Faction calls "Player Sovereignty":

> The rules of World of Warcraft are strictly designed to keep the player's actions within the expected experience engineered by the game's creators. That engineered experience however is one of war and conflict. Attempts to push the borders of acceptable gameplay are met with resistance on the part of the game creators. At best, the game is patched to remove the offending potential actions. At worst, the player is punished and their account taken away. The Terms of Service, which must be agreed to play the game, exist to control the social mores of the players. To achieve our goal, we subvert the deliberate factional conflict within the game, participating with our supposed enemies, exploring peaceful modes of play, and bringing the ideas of political player resistance to the in-game public forum. ("Third")

At the heart of such resistant play is a critique of the notion of play itself; or, as one blogger on the Faction site puts it, "I have decided to tackle the idea of [how] "play" in regards to war has become a commodity for making money in both real and virtual worlds. . . . War is a tool. Play is a tool. Together they have become a money making machine in the real and gaming world."

Other players have embraced such an ethos of resistant play, such as the Silvermoon Sisterhood, which describes itself as an "avowedly pacifist guild, pursuing ideals of non-violence in a world that crys [sic] for war and striving for peace between Horde and Alliance, outside the honorable *pas d'armes* of battlefields or any other less formal duel or tournament à *plaisance* that time and the mutual respect of players across faction might accomplish" ("Silvermoon"). Recognizing that promoting peace in a game that bases its play and rules on war is not easy, the Sisterhood provides members examples of how to use gameplay to model peace:

> To spread our message of peace, non-aggression and mutual respect to Alliance fighters, to those who cannot be reached by words as they do not share our language, Sisters must be prepared to accept death at their hands rather than run or fight back when dishonorably attacked. The recommend[ed] response to attack or threat of attack is to bow or salute and, if the assailant persists, to sit or lie upon the ground. Warlocks and Hunters must also call off their minions. ("Silvermoon")

What we see in such efforts is the application of a significant amount of critical thought—and work—aimed at examining games such as *WoW* for their ideological content. However, such work manifests not just as essays or blog postings on the game and how to play it more effectively. Rather, this critical thought is practiced and even generated within the compositional space of the game itself. The very nature of the game, with its rules and its interactivity, provides the basis upon which the Third Faction and the Silvermoon Sisterhood can practice and articulate their *dérives* through *World of Warcraft*.

GAME ON: SOME IMPLICATIONS FOR TEACHING COMPOSITION

How can we productively use students' interest in and knowledge of such games in the classroom, where many of us have to instruct students in more traditional and academic literacies? As we mentioned earlier, when we think of engaging students in discussions about technology and the new media, such as gaming, we have tended to think in terms of issues, such as gaming and violence or stereotypical representations of gender in gaming narratives and visuals. Indeed, a growing body of research is devoted to exploring the sociological dimensions of gaming (see in particular the work of Henry Jenkins), and a writing course taking as its focus such themes could easily be developed. Compositionists can certainly design rewarding courses that focus on issues in gaming, such as the vexed sociological relationship between gaming and violence or the much-needed analysis of gender and racial stereotypes in gaming narratives and visual representations. A good part of such a course could be spent analyzing and debating the many arguments put forth in academic articles examining gaming and its connection to violence, gender, and race.

While such courses may be topical and useful, we think they may miss some of the juicier thinking about literacy, collaboration, interactivity, and multimodality that can occur when we look critically at gaming with our students. More specifically, we believe that the multimodal dimensions of many MMORPGs offer a fertile ground for nurturing rich examinations of literacy—and this is ground that we must tend *with* our students. Doing so requires that we acknowledge students' extracurricular literacies and demonstrate how those literacies complement, even challenge, more traditional academic literacies. Working with gaming literacies can powerfully enhance such a project.[8] Mike's and Matt's reflections on their gaming practices and literacy development suggest some far richer uses of gaming in the writing classroom, alternatives that foreground literacy over sociological issues.

1. *Reflective and In-Progress Literacy Narratives.* At the simplest level, composition courses (in either the first-year or more

advanced upper-division sections) might take a game as the primary text for discussion, exploration, and play. Throughout the course, while learning to play the game, students can reflect on their development both as gamers and as learners working on developing new literacies. Guided reflection prompts can shape such meta-writing. For instance, the instructor might ask the following:

> What kind of composing do you find yourself doing during gameplay?
>
> What kind of composing do you wish you could do to facilitate your gameplay?
>
> What kinds of composing are more appropriate/less appropriate as you play?
>
> What's the relationship between visuals and textuality in gameplay?
>
> Describe a situation in which you could not communicate effectively with others.
>
> Describe a situation in which you could.
>
> What characterized each?

The goal in such reflections is to develop a compelling literacy narrative *in progress*; for example, a powerful assignment might be to have students compose a literacy narrative not about past literacy and learning experiences but about a literacy and learning experience they are undergoing contemporaneously, such as learning the game.

As students continue to reflect on how they engage the game as literacy learners, they can begin to think, as Mike and Matt do, about potential connections between the kinds of composing they do in the game and the kinds of composing they may find themselves doing elsewhere. For instance, Mike has begun to think about the kinds of skills he is developing in communicating while playing *World of Warcraft* and leading a "party" of online players, and how such skills may be useful in the business world:

> [If] I ever wanted to pursue any kind of business career in which I have to, you know, lead any kind of project or be a manager which I have employees working for me or anything of the sort, I'll be able to draw on my experience from the game, believe it or not. I've learned how to manage people and deal with people while being strict but not overbearing or mean or anything. I know how to deliver my points and my ideas without making people feel forced. I try to always explain why we do things a certain way.

Such rhetorical savvy and awareness of transferability of communication skills can be profitably noted in composition courses, and we can encourage students to continue to develop such connections and reflect on them. More provocatively, reflection on such experiences should point out the ways in which composing in and for game environments is rhetorically attuned and often highly collaborative— a kind of composing that is often missing in many composition courses, which focus on the individual efforts of individual students, and in which peer review is often a form of "checking" as opposed to collaborative thinking. Attention to gaming and the collective development of literacy practices it often fosters might challenge our composition courses to explore more collaborative forms of authoring and composing.

2. *Writing at Odds: Parallel Literacies and Miscommunication.* By extension, we have frequently asked students to interview professionals in different fields to find out specifically what kinds of writing those professionals find themselves doing on a regular basis. Students often discover that business professionals frequently collaborate on a proposal or business case, and much of the initial writing occurs in bursts over email or even in IM chats. Interestingly, MMORPG gamers find themselves coordinating complex gaming strategies via IM or chats embedded in the game, and such strategizing occasionally finds itself extended, as with Mike's website and discus-

sion boards, to forums outside the game. As such, the text-based brainstorming and drafting in one environment find curious parallels in another. Students can compose rich essays that reflect simultaneously on the interviews they have conducted, the games they have played, and the potential literacy skills connections between different writing environments. A particularly useful and more extended assignment emerging from such reflection might be to have students examine in depth a specific instance of unsuccessful communication during gameplay. Why, how, and in what specific rhetorical contexts did the miscommunication take place, and how was it resolved, if it was resolved? Mike's and Matt's story of the young gamer from Quebec offers a rich example of how unsuccessful communication during gameplay opens an opportunity to discuss the rhetorical significance of audience and ethos. How one uses language to craft a character, interact with others, and navigate relationships is of the utmost importance in gameplay. Critical examinations of such communications offer a nearly unparalleled opportunity to see communication move rhetorically. More pressingly, given the rise of telephonics in online gaming, more and more in-game communication is taking place orally. How do tone of voice, accent, and hearing multiple voices at once impact gameplay, and what kinds of skills and strategies do players have to develop in order to both listen and be heard? Reflecting on similarities and differences in written and oral forms of in-game communication may offer a fertile field for developing more rhetorical self-awareness.

3. *Joining the Conversation: Emic vs. Etic.* Perhaps more provocatively, we think that introducing games as texts into the writing classroom can serve as the basis for a curriculum in which the students themselves become literacy researchers. Students can, much like we have in this chapter, conduct interviews with other students interested in gaming, collect literacy narratives from gamers, and perhaps even organize focus groups for students interested in multimodal texts to discuss their

interests, their uses of such texts, and their insights into what such texts are teaching them about literacy and communication. As suggested earlier, Selfe and Hawisher's *Gaming Lives in the Twenty-First Century* already provides templates for designing interview and questionnaire protocols, and students could easily adapt and perhaps improve on their data collection techniques. Summarizing such field research—a rudimentary digital ethnography, if you will—can serve as the basis for sophisticated comparisons between what *students* say and what scholars and educators such as Kress and Gee say *about* the students and their new literacy practices. We can think of few better ways of respecting student literacies than by inviting them to participate in research and to join the conversation about gaming and literacy that is taking shape in the academy. Such a course might be not only intellectually rigorous but also illuminating for students and instructors alike, as both develop deeper insights into the impact of gaming on literacy. Having students read Gee's *What Video Games Have to Teach Us about Learning and Literacy* might be productive in a number of other ways as well. As students support, contest, or modify some of Gee's claims based on reflections on their own experience, not just Gee's relatively secondhand knowledge of gameplay, they have the opportunity to "take back" their representation, critique how others view their literacy practices, and potentially come to understand and articulate insights about such practices that we, even as scholars and teachers, cannot see ourselves. Indeed, students should be encouraged to develop a critical stance vis-à-vis gaming, perhaps augmenting and developing some of their own critical reflections about the games they play. Many popular works about video and computer gaming can be held up for critical examination. For instance, Brad King and John Borland's *Dungeons and Dreamers: The Rise of Computer Game Culture from Geek to Chic* is very useful for interrogating claims the authors make about video and computer games. King and Borland claim "computer games as a sweeping, socializing force" (7)—and not a negative, violence-inducing one, either. Specifically, they argue that "for millions of people, computer games have provided an opportunity to find other people who share similar backgrounds, stories, hopes, and dreams" (8). Other claims, however, such as that "virtual worlds are now just an extension of the real world" (8), are in desperate need of nuance: what, exactly, is extended from one "world" to the next, and is the "real world" perhaps its own extension of the "virtual world"? More specifically, and critically, whose "worlds" are we talking about, in terms of both the "real" and the "virtual"? In many ways, Mike and Matt already think about such questions, so we do not want to suggest that we bring such critical awareness to them and others potentially like them; as we have seen, they reflect on their own about the economics involved in maintaining the computer game industry, as well as some of the underlying cultural issues that designers use (consciously or not) to create their games' narrations. Rather, our role as instructors can be to help students continue to develop their critical skills, putting multiple texts, ideas, and viewpoints into play for them and helping them navigate among them and negotiate with them a reflection on their own experiences with gaming technologies.

4. *Game Design: Embracing Multimedia.* Students should also have the opportunity to experiment multimodally with their own representation as gamers. We have considered working with students to create an exhibit of photographs, videos, and literacy narratives specifically about gaming; this project could easily become a powerful and rich one for any number of composition courses. More important, such exhibits, perhaps based on the Laguna Art Museum's "*WoW*: Emergent Media Phenomenon" discussed earlier, could put into conversation different modalities of knowing and critiquing gameplay and the use of games as composing spaces. Even just recording gameplay for discussion and commentary, or documenting evidence of resistant gameplay, offers opportunities for individual and group consideration of how games

can become spaces for composing differently, and for using the tools of the game to reenvision play, interactivity, and even tactical resistances to cultural norms.

Much less simply, but much more rewarding, students can examine in more detail the kinds of cultural and ideological assumptions implicit in games by having them, as Lieberman suggests, create their own games. Such a task necessarily expands "composition" beyond the composed and beyond the traditional essay. Some initial preparation here is helpful. We have had students, for instance, read and comment on Tom Henthorne's essay "Cyber-Utopias: The Politics and Ideology of Computer Games," in which Henthorne examines computer games from the perspective of their potential utopian qualities. He notes the limitations of thinking about computer games in utopian terms: "Cyber-utopias do not allow free play. . . . The parameters of the personal utopias player[s] can create . . . are limited by the games' structures, structures that reflect the beliefs and values of the games' designers just as much as *The Republic* reflects Plato's" (64). With such ideas in mind, we can ask students to examine critically how games narrate worlds into virtual being, what assumptions underlie such narrations, and what possibilities and limitations are enabled by such assumptions and virtual world-crafting. Inviting students to compose along such lines should prompt them to reflect all the more critically about how we individually and collectively narrate worldviews—and how such narrations involve complex literacy movements that both open up and foreclose on socially desirable possibilities. At this point, a useful collaborative assignment has students design their own MMORPG or simple game, which can be done in writing, as a website, or through a simple game-authoring and -designing program (such as Game Creator). Students can reflect on the "values" they would want their game to have and think critically about how they can set up both the rules of their game and the parameters of interaction, enabled and facilitated by communications technologies, to foster those values. Comparing their own games with those such as *World of Warcraft*, in which technology is actually used to *prevent* communication between players of different "races," might reveal how communications technologies can be used to foster—or stymie—interchanges, exchanges, and understanding. This knowledge should enhance their sense of the particular rhetorical—and ideological—affordances and uses of the technologies they use.

Indeed, we encourage the composition of such games through computer media. In the process of composing games, particularly if they are composed on the Web or through a simple game-making program, students have the opportunity to develop simultaneously both technical and critical thinking skills. But moreover, working with the possibilities and limits of a program allows students to consider the particular rhetorical affordances of such programs. As we discussed earlier, Alexander and Losh describe a course in which students were asked not only to write about games but also to compose a game with a simple program. The difficulties of undertaking such an assignment are not small. Alexander and Losh maintain that the

> procedural literacy requirement [to make an executable game] may have compounded the difficulty of an assignment that already required considerable visual literacy, since they were expected to have "excellent use of screenshots / artwork," and verbal literacy, since they were charged to use "sophisticated sentences effectively," chose "words aptly," observe the "conventions of written English and manuscript format," and display "excellent document structure." Nonetheless, the presence of such an assignment in earlier iterations of the course suggests that it is a logical "next step" in developing and thickening students' awareness of and abilities with multi-literacies.

The use of such games as texts is not designed to promote a love of gaming or to justify its usefulness; if anything, the goal of helping students increase their reflective understanding of their literacy practices in one context is to prompt them to make connections across contexts. Further, as you can see from the sample assignments we have outlined, *multiple* literacies are developed through each composing experience. Inviting students to become increasingly aware of their complex literacy practices constitutes a significant part of our work as compositionists.

Inevitably, some instructors (or administrators) will question the attention we ask the field to pay to gaming as a literacy event and pedagogical space. Such instructors rightly ask, what is left out of our writing instruction when we turn to examining gaming? If students are designing an MMORPG, what kinds of writing or composing are they not doing? We cannot help but ask in return, *what are we already leaving out by not examining gaming as a literacy event for our students?* On the one hand, many of the assignments we suggest—writing literacy narratives, reporting on interviews, reflecting on writing and learning to write in different situations, conducting research through ethnography, articulating critiques of and arguments with scholarly and academic sources—are actually fairly typical, if not "traditional" writing assignments. Even having students design an MMORPG can be—and should be—a thoroughly "academic" exercise in the sense that students will have to conduct research, sift through experiences, make claims and support them, and consider rebuttals to their designs. All of these rhetorical activities can take place in the genres of the proposal, the literature review, the audience analysis, the position paper, and the research proposal. On the other hand, the critical difference, we believe, is that in focusing on a practice such as gaming we both honor an interest among many students and help students work in platforms that, in some cases, reflect and resemble the kinds of composing environments and situations that students might find themselves in outside our classes and beyond the academy. Moreover, our emphasis on inviting students to consider connections across composing environments cuts to the heart of what we as compositionists should do: offer our students a diversity of composing experiences and encourage them to become more conscious composers—that is, communicators who are rhetorically aware of how audience, genre, and tone actually work in a variety of rhetorical environments.

Along these lines, the incorporation of such texts and the move to re-vision composition as fully multimodal and multimediated raises issues of transferability. How might strategies and rhetorical abilities in composing multimediated compositions transfer to other courses, as well as other areas of students' engagement in literate activities? In our introduction and first chapter, we grounded our approach and advocacy for rethinking composition studies multimodally in the increasing multimodality of contemporary public spheres. We stand by that assertion but also note that, in terms of transferability from one "writing" context to another, multimodal composition has immense possibilities. Our gamer informants have already pointed out how they believe that composing in rich gaming spaces equips them with strategies for writing and composing in other contexts. More theoretically, however, emphasizing multiple modalities of compositional production should create robust contexts for metacognitive thinking about how texts are made, produced, disseminated, and circulated, as well as how they are picked up, remixed, and redistributed. For example, a recent issue of *Composition Forum* on the issue of transfer and writing focuses the field's attention more on habits of mind, dispositions, and threshold concepts that are needed to enable transfer. Drawing on the work of Mary Jo Reiff and Anis Bawarshi for their essay "The Value of Troublesome Knowledge," Linda Adler-Kassner, John Majewski, and Damian Koshnick maintain that the "development of metacognitive awareness is an important step toward [transfer], which 'depends on deliberate, mindful abstraction of skill or knowledge from one context for application to another.'" More explicitly, they argue that courses focusing on issues of genre

> are underscored by the position that if and when learners adopt the position that the study of writing involves consis-

> tent analysis of relationships between contexts, purposes, audiences, genres, and conventions and learn to conduct that analysis, they are both participating in the epistemological practices of the discipline and likely (so the theory goes) to be more adaptable writers. Since success as a writer is in part predicated on flexibility, these writers also will be more successful.

We can't think of courses that are more in line—both theoretically and pedagogically—with this theory than those that might emphasize compositional multimodality and the remediation of texts, concepts, figures, and ideas across multiple modalities and media. If anything, as Gee and our case studies point out, compositional flexibility is central to working with multimodal texts, so we believe that the possibility for transferable skills and strategies learned in robust communicative practices is high. More pragmatically, given the emphasis in many disciplines on the production of multimodal texts (poster presentations, lab reports and requirements documents with complex visuals, multimedia presentations), students' perception of transferable value is likely to be high.

Other instructors will worry that what we propose will not work for them because, quite simply, they have never played a computer game or do not have the technological expertise to work with games, much less design their own. In many ways, these are the ideal instructors for such a course—because they can learn with their students. More powerfully, many of their students might actually teach the instructors about gaming, allowing the students some opportunity to think authoritatively. Such newbie instructors have a rich opportunity to show their students how they—as seasoned, "master" learners—approach and acquire new literacies. Indeed, at this point in the development of multimedia and multimodal texts, we contend that many of our students are inevitably going to be—and will continue to be—ahead of us in technical capacity. Nonetheless, we have rhetorical knowledge and sensitivity that they can use to understand those texts in more complex and compelling ways. A willingness to introduce and invite such texts into the classroom may lead more profoundly and productively to collaboration and mutual learning between teachers and students.

Clearly, more work needs to be done on examining a wider range of gamers and their literacy practices, as well as the potentially specific literacy interactions and engagements of people of different backgrounds. Young women gamers, for instance, may generally favor different sets of practices (or even games), and noting such differences may be crucial to effectively designing pedagogical spaces that use video and computer games to teach writing.[9] Additionally, we need to continue to pose questions about issues of access to computer and communications technologies in general, as well as questions about the materiality of the technologies that are being engaged. How does ideology mask itself in choices of particular raced/classed/gendered avatars, and what are the effects of that masking? How do different types and levels of access to technology impact access to multimodal literacies? How do engagements with the virtual worlds of video and computer games affect our perception of the material world? These are complex theoretical questions—but they are questions that deserve attention for answers that might alter how we view the potential interaction between gaming, literacy, and pedagogy. More practically, part of the challenge we as instructors may face in thinking about gaming and writing pedagogy is our lack of familiarity with the literacy venues that students are using. Studying how "older" generations approach and engage such technologies might itself offer fruitful insights into both the technologies and the literacy practices they potentially enable.

Ultimately, we think the answers to most questions about gaming and composing may lie in paying closer attention to those already actively involved with and engaged in the new media: our students. Paying attention means acknowledging, critiquing, and taking advantage of the rhetorical capabilities of computer and online games as interactive, collaborative, and compositional spaces. Therefore, we maintain the necessity of both paying attention to, in Kress's words, "the young on the grounds of their experience" (*Literacy* 175) and offering them some of the literacy skills and strategies that we, as academic readers and writers of longer printed texts, have developed, with the goal of promoting reflection on *all* of our literacy practices. Such attention should make for richer worlds—virtual and otherwise.

5

Theorizing the Multimodal Subject

IN PREVIOUS CHAPTERS, WE EXAMINED specific kinds of multimodal composition—video production, photo manipulation, and the collaborative and interactive modalities of gaming—that challenge us to rethink what we mean by composing. Such a call is not new. We hope, however, that we have offered a strong sense of thinking that challenges both historically and contextually. We mean to evoke both the long development of rhetorical affordances of particular media as well as the rich specificity of particular moments and the rhetorical demands and possibilities that arise to meet them. As our students work more and more multimodally, examining such rhetorical dimensions allows us to see how the representation, perhaps even the construction, of subjectivity is changing. At stake, then, is an understanding of the historicity of subjectivity itself. Offering such a view, we contend, gestures toward some of the complexity of engaging and composing multimedia.

Certainly some of our students already think about such things. One immensely popular student-produced video available through YouTube, Mark Leung's "College Saga" (www.youtube.com/watch?v=wwLrgxtALWs), uses the narrative structure of the widely played *Final Fantasy* games to describe a day in the life of a college student and his friends. Ostensibly battling a force trying to convert everyone to vegetarianism, the "movie" is more noticeable for its use of an RPG (role-playing game) interface to coordinate characters as they move through the loose plot structure, much as in a real role-playing game. This video, its four parts constituting forty-plus minutes of digital "footage," spoofs simultaneously both college life and video gaming, making some telling comparisons between the hoops one has to jump through, virtually in games and figuratively in real life. The effect is ultimately parodic, at times even satirical. "College Saga" shows us that our students already contend with the multimediated self—at least creatively as they remix texts, combine narrative genres, and experiment with techno-authoring tools. What strikes us about this video is its filtering of subjectivity through the narrative of *Final Fantasy*. The video suggests much about how the narratives of such games, including their stylistic dimensions and interfaces, provide frames through which these young people articulate their own "story of the self."[1] Videos such as "College Saga" suggest that some students think about games not just in metaphorical dimensions ("going to college is a lot like gaming") but also in potentially *metonymic* dimensions ("my avatar battles its way through college"). In this case, the avatar suggests an emerging consciousness of a multimediated self, a hybrid identity, a combining of subjectivity and technology. Turning our attention, and theirs, to a more acute analysis of the kinds of subjectivities "at play" in these spaces will help all of us better understand our developing relationships with technology, as well as how we compose and author our lives, individually and collectively.

We can't look at "College Saga" and not think of Donna J. Haraway's cyborgs, especially when she writes that

> late twentieth-century machines have made thoroughly ambiguous the difference between natural and artificial, mind and body, self-developing and externally designed, and many other distinctions that used to apply to organisms and machines. Our machines are disturbingly lively and we ourselves frighteningly inert. (152)

In the midst of this ambiguity, however, we see potential. As we contemplate the dissolving boundaries between technology and humanity, "life" and "inertia," and as we watch "College Saga" and study MMORPGs, we find a rich opportunity to think about how we use such technologies, not just to enhance our communications skills, with useful literacies transferring from one mode or medium to another, but also how our sense of self changes in relation to our machines.

What kind of self is it? What is it capable of doing—and of not doing? How does it extend our sense of the humanly possible, and how also might it foreclose on other possibilities of the human? Raúl Sánchez, in *The Function of Theory in Composition Studies*, argues that "the writing subject itself . . . has remained relatively untouched, untheorized. And the result has been the persistence of a theoretical disposition that continues to understand the act of writing as an individual's rendering of experience in graphic form" (95). While Sánchez does not write at length about networked environments or multimodality, his point is particularly salient for questions of subjectivity as constructed and mediated in multimodal environments. Individuals might contribute to blogs, wikis, discussion boards, and other online forums; however, we wonder to what extent their contributions represent the voices of independent agents, speaking from personal experience, seeking active and equal participation in a multimediated public sphere. Is that sphere a Habermasian one of rational actors? Or one along the lines described by Michael Warner, of poetic iteration? Or something else entirely?

Consumer cultures only complicate matters. As some media theorists have noted, the increasing corporatization of online spaces may shift our sense of to whom we should be paying attention. In *Living Room: Teaching Public Writing in a Privatized World*, Nancy Welch notes a particular problem with "ethos" in contemporary public discourse mediated through mass media:

> Current media arrangements predispose us to define very narrowly what constitutes an authoritative appeal. Moreover, given that at least 40 percent of what passes as news on any given day is actually generated by the public relations industry for specific corporate and political interests, we face the particular problem of a thoroughly neoliberalized ethos under which the very authority to speak, argue, and act in the public realm is repackaged as a private and exclusive commodity. (135)

Gone are the halcyon days in which the Web and the Internet were lauded as new frontiers of mass and democratized participation.

Instead, we have "attention economies" in which corporate media machines stylize messages for maximum rhetorical effectiveness—measured in "hits," or how much attention is garnered by any one site, news feed, video, or posting. Richard A. Lanham summarizes some of the "rules" of the new attention economy, noting that successful practitioners should "draw . . . inspiration from your audience not your muse. And keep in touch with that audience. The customer is always right. No Olympian artistic ego need apply" (53–54). Curiously, or perhaps not at this point given our argument throughout this book, Lanham situates the development of such a view in a history of art and mediation stretching at least from Marcel Duchamp through Andy Warhol, noting the steady shift in attention in consumer culture from substance to style.

Implicit here is also a comment about authorship, agency, and composition. How does one compose effectively in multimodal environments? What is persuasive engagement in the multimediated public sphere? Effective engagement sometimes proceeds almost purely through audience manipulation, not from individual conviction. On the one hand, we might concede that this is hardly new; the ancient Sophists at times privileged rhetorical force over Platonic ideal. On the other hand, the emphasis on audience attention in contemporary multimedia settings suggests a different way of understanding compositional agency: a *corporate* understanding. We began this book with comments from Michael Warner about the increasing importance of iterability and citation in contemporary public spheres. In the constant onslaught of the new, we should not overlook the sheer rhetorical effectiveness of repetition, the comfort of the repeated, the persuasive force of stating again and again something believed to be true. In such an economy, the proliferation of individual voices might at times give way to a safe sameness, the desire for "corporate" expression, as opposed to individuated agency. What is the effect of such a possible rhetorical disposition on the sense of the composer? Sánchez suggests that "if we can account for writing without recourse to the subject, we will come closer to describing its proliferation and circulation in an increasingly networked world" (86). Perhaps. What we want to argue in this chapter, however, is that the "proliferation and circulation" of "writing" in that networked sphere has implications for the subject and our sense of subjectivity that we should not ignore.

In this chapter, we turn our attention to how we might use multimodality and multimedia in the composition classroom to teach what we call the "thick network"—that is, the many ways in which the consumption and production of multimedia texts impacts how we conceive and understand contemporary subjectivity. This task is a complicated one, we admit, and we can only catch a glimpse of the complex cyborgian intersections of media, textuality, and subjectivity. Moving toward such a view, however, even if just trying to catch a glimpse of it, seems to us a crucial part of the pedagogical enterprise of working with the many media that surround and saturate our daily lives. In previous chapters, we gestured toward how particular media and modalities—video, photo manipulation, and collaborative gaming—might be used to equip students with a greater awareness of the rhetorical capabilities of working with those media in different public spheres. Now we attempt to work multiple media together, examining a major media event, to offer a more holistic pedagogical approach focused on how persuasion works in multimediated public spheres. In service of this argument, we focus on a particular moment, examining a variety of multimedia responses to the terrible shootings at Virginia Tech in 2007. In the process, we attempt to model a pedagogy that honors the diversity of multimedia production while turning a critical eye toward such production as well.

A PARTICULAR MOMENT

On the morning of April 16, 2007, Seung-Hui Cho killed Emily Hilscher and Ryan Clark in a dormitory on the Virginia Tech campus; two hours later, he chained the doors of Norris Hall shut and began a deadly assault that left thirty more people dead and many wounded. Mixed into this tragedy were stories of heroism—engineering professor Liviu Librescu holding his classroom door shut while his students escaped through a window (and while Cho shot him, multiply and fatally, through the door); professors Kevin

Granata and Jocelyne Couture-Nowak dying under similar circumstances; students protecting other students with their own bodies—as well as bizarrely twenty-first-century twists: Cho's homemade videos sent to NBC, his classroom writings posted on the Web, a variety of video commentaries on his mental state (and classroom performance) on news sites like cnn.com, and, on listservs, blogs, and wikis, an outburst of professional discussions about Cho, his writing, and the institutional inability to prevent the killings.

Many of us received the majority of our information about the murders through the intense, probing televisual coverage available on major news channels. At the same time, the significant amount of online chatter and discussion about the events marked a potential turning point in how many of us in this country receive and respond to news. Indeed, while televisual and techno-response forms a significant dimension of many such tragedies,[2] what drew us to the Virginia Tech shootings in particular were (1) the intensity of textual and visual dissemination of the tragedy through online forums and (2) the intensity of response in online discussion of the tragedy. Granted, Columbine shooters Eric Harris and Dylan Klebold may have "documented their rage on home video" in 1999,[3] but five years later, Cho's documentation of his rage could be almost instantaneously disseminated and made subject to widespread commentary through blogs and YouTube. Moreover, the transfer of the burden of truth-telling from Cho himself to the body of his texts further distinguishes this tragedy from others. A variety of texts—from public statements made by university officials, to newspaper editorials, to video interviews of those on campus during the shootings, to commentaries by administrators and mental health officials—emerged in the immediate aftermath of the murders and Cho's suicide. Prominent among such texts were Cho's own videos and plays, which drew considerable and oddly pedagogical commentary ("Cho"). Whence came this need to discipline Cho's writing? As Matthew J. Newcomb notes in "Taking Attendance: Absent Writing and the Value of Suffering," writing about pain does not mean "connecting everyone to a universal sense of loss, nor is it simply about finding ways to fill up the space of loss.

Instead, writing about suffering involves recontextualizing absences so that they will have direct uses for other specific groups" (776). One consequence of Cho's absence as author, then, was that his writing became ever more present, and the ravenous critique of it was our recontextualization of his absence, part of our need to "create structures of things that are directly sayable" (Newcomb 776).

The Virginia Tech shootings therefore provide a compelling case for examining the complex intersections of new media and subjectivity. At the same time as editorials and short essays, online and in print, offered interpretations and commentary from an ever-increasing distance, a simultaneous new media response emerged, a dizzying accumulation of blog postings that attempted to make sense of—to account for—the violence and its perpetrator. Such postings seem to us emblematic of a textual production combining both speed of response and disciplining of affect. This particular constellation of new media texts surrounding Seung-Hui Cho's horrific act serve as the impetus for our exploration of technology and subjectivity here. In this chapter, we consider the conflation of writing with normalizing knowledge in the "aftermath texts," and the speed with which such conflation was put in the service of disciplining responses to the tragedy; by doing so, we hope to link the aftermath production of rich, varied, and, at times, disturbing content by everyday Internet users to larger conversations about subjectivity. Our central question: how can we use new media to open up spaces, not just for immediate response but also for critical reflection?

We offer this analysis uneasily, recognizing that we must consider the events and the production of texts in their aftermath with great care, ever mindful that such consideration might be considered exploitative of the tragedy of the murders. At the same time, we believe with Richard E. Miller that we need to find ways to provide our students (and ourselves) "with concrete opportunities to engage in the creative work of generating local, temporary solutions in an imperfect world" (37). This essay gestures toward that goal; we believe that careful and critical attention to such events and texts is necessary lest respectful silences that often accompany tragedy become normative silences precluding much-needed discussion.

THE SPEED OF NORMALIZATION

In many ways, Cho was a highly literate technology user, making use of the Internet to obtain guns and ammunition clips, as well as to create content for dissemination (primarily videos explaining why he did what he did). That content, made available through a variety of news sources and on YouTube, inspired others to create and distribute their own response videos and other texts, which in turn generated even more online commentary and discussion. We do not take up the all-too-daunting task of explicating Cho's videos; rather, we focus on the rapid production of and participation in new media texts surrounding the murders at Virginia Tech and also Cho's own writing. This phenomenon offers insights into new media subjectivities, laying bare the simultaneous opening up and closing off of what Haraway has famously referred to as "spaces and identities and . . . boundaries in the personal body and in the body politic" (170).

Almost immediately after the shootings, two of Cho's plays, *Richard McBeef* and *Mr. Brownstone*, were posted on News Bloggers, a now-defunct AOL site that billed itself as "hard news, raw opinions, penetrating perspectives" ("Cho"). The plays generated their own share of intense discussion and debate and were made available by one of Cho's fellow playwriting students, who described a peer workshop session thus:

> When we read Cho's plays, it was like something out of a nightmare. The plays had really twisted, macabre violence that used weapons I wouldn't have even thought of. Before Cho got to class that day, we students were talking to each other with serious worry about whether he could be a school shooter. I was even thinking of scenarios of what I would do in case he did come in with a gun, I was that freaked out about him. When the students gave reviews of his play in class, we were very careful with our words in case he decided to snap. Even the professor didn't pressure him to give closing comments. ("Cho")

Note how the description of the plays—"twisted, macabre violence"—immediately shifts to an imagining of what Cho must be like, what his subjective experience of the world might entail: "he could be a school shooter." In that light, the following carefully worded comments are all the more intriguing:

> While I was hesitant at first to release these plays (because I didn't know if there are laws against it), I had to put myself in the shoes of the average person researching this situation. I'd want to know everything I could about the killer to figure out what could drive a person to do something like this and hopefully prevent it in the future. Also, I hope this might help people start caring about others more no matter how weird they might seem, because if this was some kind of cry for attention, then he should have gotten it a long time ago. ("Cho")

For this student, the plays give clues about the "killer" and "what could drive a person to do something like this," with the goal, of course, of preventing it in the future. However, the rhetoric of prevention gestures powerfully toward normalizing subjectivity (the "shoes of the average person"), and the student reads the plays as a posthumous confession of a tortured soul. More specifically, the comments collapse Cho's writing and his subjectivity: the killer wrote plays about killing and the playwright interested in writing about killing is a killer.

Clearly, Cho was troubled. It is worrisome, however, to see how writing itself is figured in these discussions. It is a figuration immediately picked up on by numerous blog postings written in response to the plays:

> How was he not kicked out of school for this? My alma mater would not have permitted it. I can't even believe this wasn't turned into the counselors.

> Um . . . never mind the fact that it's written like, oh maybe a 6th grader who thinks it's cool to swear, the bigger question is, why wasn't something like this brought to the Chairperson of the Arts Department or even the Dean. Clearly . . . there

> is an undeterminable amount of rage here. Now . . . it's too late to do anything. My heart breaks for those kids and their families. ("Cho")

Increasingly, posts in this series focused on the issue of responsibility, furthering the sense that the writing should have been read by authorities as a signal of imminent catastrophe:

> This guy acted out like this and nobody said anything to the authorities or the teachers? Wow. How many Columbines do we need before somone figures this out?
>
> Wow. Why didnt the teachers or counselors do something? This writing is obviously evidence of someone really disturbed. This tragedy might have been prevented if only someone reached out to this twisted soul.
>
> wow, and he didn't get checked out after this? i cant believe noone said anything.
>
> Can anyone confirm the mental capacity of this student? For a college senior- his uses of "name-calling" and word choices are completely incohesive and give a sense that this student may have been mentally challenged. It almost sounds like what you might here from people undergoing psychoanalysis to deal with physical/sexual abuse and reassume their childhood personalities! ("Cho")

In some posts, the "analysis" of the relationship between Cho's writing and his suicidal rampage gets caught up in narratives skeptical of the efficacy of education and literacy instruction:

> This is garbage, a Senior in College wrote this? This guy had obvious problems, and from what I understand he was referred to a mental health clinic on campus. Guess what: YOU DROPPED THE BALL. But you already know that by now. ("Cho")

The dropped ball in this case seems both the inability of psychological systems to recognize someone in need of intervention and the failure of literacy educators to pinpoint and address Cho's "twisted soul." The failure to catch Cho through his writing seems to be cast as the failure of literacy education more broadly to understand writing as a mode of disciplining subjectivity. Indeed, inasmuch as such comments dominate the blogs, they create a narrative of Cho's subjectivity that demands not only a psychological accounting (how could he have done it?) but also a panoptical accounting (why didn't anyone see it coming?). We see here a longing for the "grid of control" that Haraway refers to in her "Cyborg Manifesto"—one disturbing possibility within a cyborg world.

As a critical intervention in the discussion of this material, we would like to offer an analysis of both how quickly the postings were disseminated and how the speed of production of the content canted the discussion toward discourses of subjective normalization. We engage here the thinking of both Michel Foucault and Paul Virilio: Foucault for his discussion of the production of normalized subjects through disciplinary regimes and Virilio for his account of technology and its impact on subjectivity. In *The Information Bomb*, Virilio revisits McLuhan in order to claim that it "is not the medium which is the message, but merely the velocity of the medium" that is significant in our technologized experience of information (141). The speed of dissemination and response creates a situation in which we access affect at the expense of understanding: "Digital messages and images matter less than their instantaneous delivery; the 'shock effect' always wins out over the consideration of the informational content" (Virilio 143). The subject—the thinking agent—can disappear into automated "feeling" about information; in the Virginia Tech case, the speed of dissemination of and response to Cho and his writing may have helped us feel certain ways, but it did not help us think. The speed of dissemination leads to a potential "automation of response," a closing down of critical consciousness in the quick dissemination of "information":

> We gain information instantly at the cost of becoming information ourselves, outside of any consideration of personal choice, as liberal political theory understood it. New possibilities open up to us, but only as they become technologically

> efficient, manageable and therefore standardized. The horizons of the subject are simultaneously expanded and reduced. (Mansfield 155)

The increasing rapidity of information exchange interpolates subjectivity in terms of both trajectory and interiority. Our "horizons" may be broadened by the information we receive, but its velocity beggars our ability to make sense of it. In that impoverishment lies our own beggaring, as our ideas, insights, and communications become part of the rapid and rapidly beggared content of the networks. For Virilio, the political implications are clear:

> Ultimately, it is the whole problem of immediacy and instantaneity in politics that is posed today. After the authority of human beings over their history, are we going to yield, with the acceleration of the real, to the authority of machines and those who programme them? Shall we see the mechanical transference of the power of the political parties to the power of electronic or other devices? (122)

We acknowledge with Virilio a shared sense of loss of control in information management, even in "self-management," as we see how new media—in this case, the Internet—monitors and disseminates information about identities, specific and collective. At the same time, we want to thicken Virilio's description of networked subjectivity. After all, while programmers can capture and manipulate information we upload to the Internet as one form of control, the velocity of information itself disciplines in more covert ways. That is, in Virilio's accounting of the production of "technologically efficient, manageable and therefore standardized" subjectivities, we can also read Foucault's understanding of the production of normalized subjects. For Foucault, normalizing narratives circulate around and through us, conditioning our subjectivity, sculpting our sense of what is right, proper, appropriate, or "normal." In *Discipline and Punish*, Foucault offers the concept of the "normalizing gaze," or "a surveillance that makes it possible to qualify, to classify and to punish" (184). We are all subject to this gaze, our behaviors offered up for review and discipline by others (and, ultimately, by ourselves). To what extent—and how—do technologies of communication further the range of this gaze?

Bloggers and respondents do not subject only Cho—the anti-model of the normal—to analysis; they also discipline one another with the same velocity of response. Note the replies to Stacy, who comments on Cho's writing ability:

> 1. This is something written by a senior in college? It reads and sounds like something a 9th grader might write.
> Stacy at 2:28PM on Apr 17th 2007
>
> 2. Hey Stacy, how about you keep your rude comments to yourself. The point of their story wasn't supposed to be so you could talk about their writing. You're supposed to reflect on the events that took place yesterday and realize what a tragedy it is. So many people lost people they loved or had people hurt because of some psychotic persons selfish acts.
> Madeline at 2:37PM on Apr 17th 2007

Quickly, the comments turn from admonishment to outright hostility:

> 3. You are the one that sounds like an idiot: Get a life Stacy.
> MonMon at 2:39PM on Apr 17th 2007
>
> 4. Stacy,
> I don't even know how to comment on what you just wrote.
> Maybe, grow up!!
> Roxie at 2:40PM on Apr 17th 2007
>
> 5. Hey you twit - - - we aren't reading it for correct grammar or English. It is written about his twisted mind. Get focused moron. . . :o(("Cho")

The time stamps indicate that other Web users discipline Stacy, almost immediately, to pay attention to "really" important issues. It is not the writing itself but its narration of Cho's deviance, his outlaw subjectivity itself, that is significant as knowledge, and hence power. The idea of the transparency of writing—of narratives that provide us a clear view of the deviant underneath—operates powerfully within this structure. This disciplining comforts us, as Lynn Worsham writes in "Going Postal: Pedagogic Violence and the Schooling of Emotion," because it offers us "the view that violence is the unfortunate result of individual pathology" rather than an outlaw response to regimes of affect that are the "primary and most valuable product" of late consumer capitalism (219). This disciplining tells us: If we just read the situation right, we can prevent it from happening again.

Certainly, not everyone responded the same way.[4] However, the call to make the writing confess, and for us to learn to spot such a connection, so saturated the media that the *Chronicle of Higher Education* devoted a column to the issue, pulling together a variety of postings from the Web examining the connection being made between Cho's creative writing and his violence. Most respondents argued that we cannot use such writing as an assessment of normativity:

> *Edward Falco, Virginia Tech*: There was violence in Cho's writing—but there is a huge difference between writing about violence and behaving violently. We could not have known what he would do.
>
> *Stephen King*: On the whole, I don't think you can pick these guys out based on their work, unless you look for violence unenlivened by any real talent.
>
> *Tyler Dilts, California State University at Long Beach*: The question we need to ask ourselves and each other is whether writing about issues and experiences that illuminate the darker parts of human experience is a worthwhile endeavor. I think if we are honest in our answers, we must say that it is. How can we suggest that we should limit our students' freedom to write about the stuff of life, which we claim informs great literature. ("Creative")

Such comments help to complicate the discussion of the tragedy, creating resistant readings of the situation that refract the crystallization of normalizing discourses around it. If we expand our view to see the postings not just about these plays but also about the tragedy in general, the processes of subjective normalization are much more diffuse. One of Jonathan's former students, who has shared with Jonathan his own posted content on YouTube, "favorited" a video response to the video Cho sent to NBC. Jonathan followed the link to Cho's own horrifying video and noticed multiple video responses, as well as numerous message board postings. As of last count, each video response had been viewed nearly 1,000 times each. Comments proliferate, some highly thoughtful, some extraordinarily crass. Some postings are simply finger gestures, commenting on others' postings:

```
... ... ... ... ../'¯/) ... ... ... ...... (\¯`\
... ... ... ... / ... .// ... ... ... ... ..\\ ... .\
... ... ... ../ ... .// ... ... ... ... ... .\\ ... .\
... ../'¯/ ... ./'¯\ ... ... ... .../¯`\ ... .\¯`\
../../ ... / ... ./ ... ./.|_ ... ... _|.\ ... .\ ... .\ ... \.\..
(.( ... .( ... .( ... ./.)..).. ..(..(.\ ... .) ... .) ... .).)
.\ ... ... ... ... ... .\/ ... / ... .\ ... \/ ... ... ... ... ... /
..\ ... ... ... ... ... .. / ... ... ..\ ... ... ... ... ... ... /
... .\ ... ... ... ... ..( ... ... ... ... ) ... ... ... ... ../
... ... \ ... ... ... ... .\ ... ... ... ../ ... ... ... ... ./
```

The extremity of the situation, coupled with the ability to post text and video responses nearly simultaneously, probably accounts for much of the unreflective, knee-jerk, even reactionary commentary.

While such proliferating responses might diffuse the normalizing gestures we encountered earlier, in other ways the normalizing practice continues. For instance, one YouTube poster, ChoLovingVampire, posted a video montage of Cho's original videos entitled "The Warrior Seung Hui Cho," to which he appended the following note: "This is a dedication to Seung Hui Cho because I can relate to him. Please leave comments." As you might imagine, the responses are varied, with many lashing out at ChoLovingVampire for daring to sympathize, much less potentially empathize with Cho. How might we understand such a video?

On the one hand, "The Warrior Seung Hui Cho" attempts to initiate an alternative reading of the Virginia Tech tragedy, one that works against simply pathologizing Cho and instead toward understanding the reasons, potentially systemic reasons, for why Cho acted out so tragically. Some commentators pick up on such a sympathetic reading, posting their own comparable videos. One, SeungHuiChoHero, collects a series of Cho-related videos that foreground sympathy for the killer, creating a repository for tribute videos. A quick YouTube search leads to many others, suggesting a potential critical mass of individuals who feel sufficiently alienated from society to sympathize with Cho. As SeungHuiChoHero puts it in the "About" section of his YouTube profile: "This thing, my life, all an agony, of Hell of torture. . . . And years of bludgeoning torment tiny nuisances. The disgust eyes, dirty frowns, and red fingers pointing at me. Feeling all the patheticness and humiliation. What time is right to abort the null existence and retire from sick lifeblood."

On the other hand, the construction of the videos themselves is pretty brute and minimalist, focusing on simple montage and pounding heavy metal music; in the case of ChoLovingVampire's video, the music is Disturbed's "Indestructible":

> Indestructible
> Determination that is incorruptible
> From the other side
> A terror to behold
> Annihilation will be unavoidable
> Every broken enemy will know
> That their opponent had to be invincible
> Take a last look around while you're alive
> I'm an indestructible Master of War!

The video producers compile images of Cho and sound recordings from his statements, sometimes accompanied by subtitles so we can understand what he says. The ultimate effect is one of video homage focusing on personal identification with Cho, as we can hear in the pathos-laden commentary from SeungHuiChoHero.

Curiously, these video responses come two to three years after the shooting, suggesting little advance in the critical discussion of the tragedy in the loose public sphere of YouTube.

The homage reaches particular intensity when another video tribute, ThePelly's "KEKEKE—The Ballad of Cho Seung-Hui" (www.youtube.com/watch?v=NeF1t07s-GA), made with an identical aesthetic to ChoLovingVampire's video, features a rap song by Ryan Lambourn, who sings:

So, here we go
It's Seung-Hui Cho
Killin all the trust fund deceitful charlatans
And Yo, now you know
It's Seung-Hui Cho
Spilling the blood of a million dollar kid

You can then follow a link to a gaming site (newsground.com) where Lambourn has posted an online game in which you can pretend to be Cho yourself. "V-Tech Rampage" (at www.newgrounds.com/portal/view/378086) opens with a screen announcing: "The pawns are all in place, the time has come that i [sic] may finally send my message to all the world." Press *S* to talk, *A* to shoot, as the game leads you around a campus so that you can kill various students and other passersby. Made with a simple game design program, "V Tech Rampage" invites you to occupy a first-person shooter point of view, but to what effect? To bleed off anger and hostility? To get cyber revenge on those who immiserate you? To imagine Cho's alienation? Message board comments to the game tend to echo earlier commentary castigating any potential sympathy for Cho:

> Dude, this is sick! This was a real tragedy! V-Tech is not something to make fun of, but I guess you thought otherwise, didn't you?! Woke up one morning and thought "Know what? I'll be a complete jackass and make a game where you get to reinact [sic] the most tragic college shooting in US history." You sir are an asshole.

Other commentary lauds the "retro" design of the game, which looks like an old Atari game from the late 1970s and early '80s: "This game is so much fun! I like the retro style with the retro music and all, the automatic fire is great, the AI is great, unlimited ammo is also awesome. FTW ["For The Win"]."

The transformation of the original shooting into binary-driven message board commentary, video homage, and even a first-person shooter offers a suggestive narration, not just about the tragedy but also about the *transmediation* of the tragedy. In each case, we see the potential for alternative critical commentary; Lambourn's song, for instance, opens up the possibility for an interesting class analysis, and while this reading might not excuse Cho's actions, it points to how class divides might engender damaging affects of alienation and self-hatred. At the same time, however, "V-Tech Rampage" seems to dilute such potential by reindividuating your consideration of the tragedy as you play a cartoon character shooting students; you ultimately focus on just playing a game that generates an affect of nostalgia for earlier, simpler gaming platforms as opposed to producing more complicated assessments of the tragedy depicted. Indeed, the game gestures toward an earlier "speeding through" a consideration of the original event as you move quickly through the campus, scoring points that take you nowhere.

We might argue that the extreme individualizing of responses levels any critical discussion out of existence. It is, in some ways, to "quote" an earlier comment, given the finger. In *Discipline and Punish,* Foucault suggests that our conviction that we are all truly individuals serves as one of the most powerful disciplinary legacies of the modern era; he writes that "discipline 'makes' individuals; it is the specific technique of a power that regards individuals both as objects and as instruments of its exercise" (170). Our individuation better conditions us to give an accounting (a confession) of ourselves (as measured against the ever-shifting norm); we so value the sound of our own voices that we respond to discipline, confess, point the finger. Individuation, combined with earlier speeds of response, merge in the collapse of critical distance. Such intense individuation, perhaps, results in the diffused potential for collective response, for potent clusterings that might form in order to explore, question, and pose more meaningful interpretations.

As Haraway writes, the cyborg suggests "a way out of the maze of dualisms in which we have explained our bodies and our tools to ourselves. This is a dream not of a common language, but of a powerful infidel heteroglossia" (181). However, we do not find "powerful infidel heteroglossia" in this collection of voices and divergent views. Certainly, the ways in which the voices cluster and diffuse themselves speak to the multimodality of the representation of trauma, affect, subjectivity. But the possible heteroglossia here speaks more often than not with normalizing gestures designed to discipline what multimodal composing can and cannot do; the overwhelming sense is that one's composing should be made to confess a normalized sense of self—or a nonnormalized sense of self in need of further discipline. Moreover, the intense diffusion and speedy proliferation of composing and of video and game production asserts the primacy of individuality to the near exclusion of critical purchase and intervention—not to mention collective response. Such heteroglossia seems neither powerful nor infidel, but rather both disciplined and diffused in its individuation.

TEACHING THE THICK NETWORK

Much discussion of this tragedy in the popular press, as well as in our own professional conversations on listservs, on blogs, and in campus meetings, focuses on how writing teachers might more effectively (1) diagnose "dangerous" writers before tragedy and/or (2) counsel students after tragedy (usually in the form of writing prompts that encourage students to write "through" their feelings). In either case, writing serves as a seemingly neutral affective space, a conduit equally capable of surveillance and therapy. In the aftermath of tragedy, discussions of technology happen rarely, and discussions of the potent mix of technology, writing, and emotion not at all. As Peter N. Goggin and Maureen Daly Goggin remind us in "Presence in Absence: Discourses and Teaching (in, on, and about) Trauma," such limits on discussion come at great rhetorical price:

> Political, institutional, social, and cultural (pace rhetorical) conditions permit some to speak while eclipsing others, permit some views while silencing others, and permit some forums while ignoring others. In short, who gets to speak and be heard, who has access to public forums, when and where this happens, and what can and cannot be said and heard are crucial rhetorical questions that problematize in important ways the understanding of trauma and writing (about) trauma. (33)

We want (that is, what we both desire and find missing is) a strong sense of how we discuss trauma, technology, and subjectivity with our students in ways that open up the spaces of schooled emotion. We might find such spaces far beyond our almost instinctive rush to diagnose and therapize through writing, and even further beyond the technological discussions that too often focus on the cool critical thinking skills we can develop with video games, or the easy expertise offered by iMovie software. Such spaces demand that we shift our focus to how technology and composing affect faculty and students.

We want a more (dare we say) humanistic—yet still critical—literacy of technology, one that takes as part of its ecology the affective realm of technology and technology use. As Cynthia Selfe and Gail Hawisher remind us, the "cultural ecologies" of literacy take into account the "complex web" of sociocultural factors that shape (and are shaped by) our approach to communications technologies (*Literate* 29). We believe that the construction and critique of subjectivity in technology forms one such factor. For, as Selfe and Hawisher note, "We can understand literacy as a set of practices and values only when we properly situate our studies within the context of a particular historical period, a particular cultural milieu, and a specific cluster of material conditions" (*Literate* 5). Given our field's ongoing consideration of subjectivity and the institutional site of composition (and given that technology and technological access form part of the institutional site itself); given our ongoing interest in discussions of emotion (Worsham), pathos and ethics (Micciche), and trauma and rhetoric; and given the dizzying speed with which mainstream US culture seems to shoot along the S-curve of technology at the same time that it ratchets up its disciplining of "terror" and "dissent," we firmly believe that considering subjectivity and technology together, here, at this moment, is a necessary intervention into the discussion of literacy itself.

As we (teachers and students)[5] grapple with new media and multimodality, it is all too easy to fall into instrumentalism and ignore, even briefly, how that media form part of our very sense of being, part of the material grain of who we are and how we interface with the world around us. Sheridan, Ridolfo, and Michel argue forcefully in "The Available Means of Persuasion: Mapping a Theory and Pedagogy of Multimodal Public Rhetoric" that

> it might have been possible for rhetorical education to overlook materiality in a writing-centered culture in which multimodal rhetoric was beyond the reach of nonspecialists and the tasks of distribution belonged to individuals other than composers. But our options and resources in a digital age have broadened significantly. . . . Educational sites designed to prepare students to produce multimodal public rhetoric need to help students negotiate the material processes associated with effective rhetorical intervention. (818)

We couldn't agree more, and it is here that a critical awareness of the intersections of affect and technology offers much to help us understand how we might help our students (and ourselves) "negotiate the material processes associated with effective rhetorical intervention." In "Going Postal," Worsham defines emotion as "the tight braid of affect and judgment, socially and historically constructed and bodily lived, through which the symbolic takes hold of and binds the individual, in complex and contradictory ways, to the social order and its structure of meanings" (216); we see that "tight braid" being further complicated, in this particular instance, by new media capabilities and a cultural anxiety about technological abjection. Our technological anxiety is ever-present to us in mass media cultural products: films from *Metropolis* to *A.I.*, from *Desk Set* to *The Matrix* trilogy posit individual, Western-style subjectivity as an Edenic state and technology as threateningly abject—or objectifying. In this system, emotion and desire (properly placed) become the signifier of what is human; the "tight braid of affect and judgment" that serves as the conduit for symbolic order is knotted together with our constitutive and constituted technologies. Further, the anxiety about subjectivity undergirding our approaches to technology becomes a key part of the cultural ecology of any technological literacy.

For surely, the disciplining evident in these new media responses to the Virginia Tech murders has to do with a symbolic order taking hold of and binding respondents to social order through Worsham's "tight braid" (216), a braid of affect, judgment—and multimodal technologies. Particularly in the discussions of violence, of right and wrong, of ethics, that is, we would do well to remember Laura Micciche's discussion of emotion:

> [Emotion] is crucial to how people form judgments about what constitutes appropriate action or inaction in a given situation—precisely the realm of ethics. The idea here is that emotions, like reasons, move people to judge, decide, and act in certain ways, and that, consequently, emotion is central to rhetorical action. (169)

If new media offer us new realms in and about which we must engage in such action, it also offers the possibility of new regimes of emotion, new possibilities of resistance and control, new cultural ecologies of cyber literacy. And it is crucial that we, in the business of literacy education, try to understand these possibilities. Thinkers such as Michel de Certeau and Maria Bakardjieva have provided useful methods for understanding "tactical" uses of technology in general (and communications technologies in particular). That is, these thinkers have helped us explore how "everyday users" work with new media to create new "use genres" for technologies that extend and even resist the original purposes for which the technologies were designed.[6] It's important that we, students and faculty, understand not only how to use new media technologies (to be "functionally literate" as Selber puts it in *Multiliteracies*) but also to have a strong critical sense of the discourses that enable these technologies.

Along such lines, we hope our analyses of new media postings about and responses to the Virginia Tech tragedy open a space in which to think about the interconnectedness of writing technologies and our very sense of self(ves). These analyses are hardly comprehensive, and, given the amount of material on the Web, our comments can only be suggestive. Rather, we have tried to explore a particular instance of writing/composing through new media at the same time that we foreground critical considerations of how subjects are themselves "composed." While we can offer no easy answers here, we believe that such an examination must be part of any framing of new media literacies, particularly given students' increasing ability to produce and post content. Our strong sense is that we should turn our students' attention to discussions that help link our immersion in network technologies with much-needed theoretical critiques of how those technologies intersect our subjectivity(ies).

Monica Barron points out that changes in pedagogy might have limited ability to "preempt a violent manifestation of mental illness in a school. But let's begin where we are, paying closer attention to the emotions evoked by the material we teach, by the classroom dynamics, by the work the students produce" (43). Like Selfe and Hawisher, we should invite students to tell us—and in the process tell themselves—their stories about their use of technology. We can also extend Selfe and Hawisher's one-shot surveys and ask students to keep daily writing/technology logs, to interview one another, to become aware of the "profusion of spaces and identities" that Haraway sees underlying the "network ideological image" of cyborg subjectivity (170). Perhaps we should also examine with our students some of the Pew reports about the use of the Internet, particularly young people's use of technology—how do they position young people: as consumers? agents? tacticians? We might at least prompt students to contrast their figuring as young people with their own auto-ethnographic reflections on their experience of the Internet. Drawing from ethnographic methods, for example, Selber asks students "to study and produce thick descriptions of the social conventions in actual computer-mediated communications" (*Multiliteracies* 54). Similarly, Amy Devitt, Anis Bawarshi, and Mary Jo Reiff advocate a blend of genre analysis and ethnography to encourage students to "recognize how 'lived textualities' interact with and transform 'lived experiences'" so that they can "participate more knowledgeably and critically in various sites of language use" (549–50).

In turn, this attention to auto-ethnography and the figuring of the self leads to greater possibilities for critical discussion of affect and new media with our students. As Devitt and colleagues point out, paying attention to text and experience shows "that the community is not just a backdrop to language behavior, but a growing, moving environment that includes texts and speech as its constituents, just as people are its members" (550). Further, as part of the two-volume *JAC* issue focused on trauma and rhetoric, Maya Socolovsky asserts that the "speed and immediacy" of online response to trauma can itself reassert human agency "into the machine," as it were. In her discussion of online memorials in the aftermath of the Columbine shootings, Socolovsky writes that the "constant vacillation" brought about in the very act of reading online allows a particular (and even beneficial) blurring of self and other: "If the body no longer serves as a limit to our subject position or experiences, then sociable interactions and acts of empathy online are mediated through a faceless and bodyless community. . . . That is, a virtual space becomes a place of belonging and even of collectivity" (469). With this sense of community in mind, students may find a more thorough critical engagement with the multimodalities of social networking particularly compelling. We could ask, for instance, how social networking sites like Facebook and Tumblr structure our experience of the Internet. Placing tentative answers to that question into conversation with accessible works such as David Brin's *The Transparent Society* could be productive. In his book, Brin, an award-winning science fiction author and technology expert, worries that the emerging culture of technological surveillance creates more opportunities for us to be monitored. What happens, we might ask our students, when they take the power of surveillance into their own hands, with a smartphone, a YouTube account, a surreptitiously made video? How does that disrupt—or does it—the overweening culture of surveillance?

In any of these approaches, we advocate making use of new media technologies themselves to have these discussions; it is important that we (1) assume and encourage our students' (and our own) greater functional literacy (à la Selber) and (2) expand our own approaches to critical literacy, moving beyond what Steve Westbrook refers to as the "consumer bias" in critical thinking approaches (460), which treats "visual artifacts as if they have always already been produced and confine the visual rhetor to the role of respondent, who uses a variety of semiotic strategies to understand the effect of the preexisting visual image" (461). Inviting students to put their logs, blogs, surveys, and interviews into conversation with critical cyberculture theorists may help us (and our students) develop intriguing insights into the relationship between technology and subjectivity. Further, as Barron notes, such moves may help us all "think constructively, as Seung-Hui Cho could not, of [ourselves] as members of communities and as producers of work that evokes emotion in others, for better or for worse" (43).

Much provocative thinking from the last several decades may provide a way into prompting students to talk to us—and to talk back to the theorists—about their use of technology. For instance, how might students respond to Virilio's often strident accounting of the technologically enabled short-circuiting of critical consciousness? We understand Virilio as speaking powerfully to the production of aftermath texts and their "immediacy and instantaneity" as problematizing, if not undermining, the "authority of human beings over their [own] history" (122). But this story is not the only one about technology and its uses, about technology and its power. If Foucault is correct in maintaining that "power is tolerable only on condition that it mask a substantial part of itself[, i]ts success is proportional to its ability to hide its own mechanisms" (*History* 86), then how might power put technology—and our combined fascination with and horror of it—to work in disseminating regimes of the normal? Our analysis of responses to Cho's plays and videos might itself serve as a "text" for classroom discussion in which students could be asked to develop resistant readings of the normalizing gestures of blog respondents; they could also work toward developing interventive strategies to engage critically the kinds of comments that quickly move to silence views in the service of promulgating standard, normative responses.

We must also become aware of how our own discipline—English studies in general and composition/rhetoric in particular—serves monitoring, disciplining, and normalizing functions within the academy. We often take pride, especially since the "social turn," in offering our students "critical pedagogies" that attune them to the manipulations of a consumerist society, that apprise them of the ideological structures that undergird capitalist economies, and that advise them to think critically about language, discourse, and power. At the same time, our courses ostensibly prepare students for participation in the very structures we ask them to critique. We serve as gatekeepers, certifying that students have certain kinds of literacy skills that will enable them to find jobs in the very society that we problematize for them. Where do we draw the line at questioning the structures, the regimes of power, the ideologies that normalize "literacy"? Tellingly, we often draw that line when we come too close to emotion, to engaging the critical powers of feeling. We often shy away from some topics, such as the Virginia Tech shootings, out of respect for those involved, but also because we fear that broaching such topics might overstep our disciplinary boundaries. We thus discipline ourselves and our students away from difficult topics and foreclose on the possibility of understanding and responding to those topics critically.

What is at stake is how we might make the dual mission of literacy education—to be functionally literate (again, in Selber's sense) and to critique—become part of the acknowledged conversation about literacy in our society that we have with our students. We might take seriously Foucault's call to rethink knowledge production: "The transformation of one's self by one's own knowledge is, I think, something rather close to the aesthetic experience" (*Ethics* 131). Even more provocatively, he argues that "the experience of the self is not a discovering of a truth hidden inside the self but an attempt to determine what one can and cannot do with one's available freedom" (276). The term *available freedom,* though, implies that there are limits, sometimes unknown limits, that structure our sense of the possible. Some of the limits might be the constraints that we place on ourselves when we refuse to broach difficult topics such as the Virginia Tech shootings and the ways we use technology to process—and normalize—our experiences, our reactions, our emotions. However, as Richard Miller argues, our job as humanities scholars and teachers

> is to establish an environment that promotes reflection and to provide our students with multiple opportunities to experience mental acts that take them to the edge of the unknown. . . . To counter the instinctive tendency to react to the unknown with preemptive judgments about others, we can illustrate the challenges and the opportunities that come from living in a multi-perspectival world. (36–37)

A key site from which we might illustrate such challenges and opportunities is the intersection of emotion, self, and new media technologies in the writing classroom. We should ask, with our students: what are the limits on our experience of freedom—freedom to discuss, to consider alternative views, to push the boundaries of the known and the knowable? To ask this question (let alone attempt to answer it), we must look critically at the emerging metanarratives of cyberculture, of our interactions with technology, of technology using us, of our becoming cyborg. We must see the creeping kinds of normalizations and disciplinary regimes that condition our experience of available freedom.

Perhaps what we need most is to step back, contemplate the speed that Virilio sees as one of the chief characteristics of our very human interaction with technology, and reflect more critically and deliberately about our mutually constitutive relationship with technology. Along such lines, we recall that the myth of the cyborg has served as a good starting metaphor for understanding and reunderstanding our relationship with technology—and our relationship to ourselves as mediated, and changed, through technology. What happens, then, when we reembody the cyborg and reinvest our own cyborged humanity with affect? Haraway herself comments on the permeability of bodies and technologies:

> Technologies and scientific discourses can be partially understood as formalizations, i.e., as frozen moments, of the fluid social interactions constituting them, but they should also be viewed as instruments for enforcing meanings. The boundary is permeable between tool and myth, instrument and concept, historical systems of social relations and historical anatomies of possible bodies, including objects of knowledge. Indeed myth and tool mutually constitute each other. (164)

We want to turn our attention to the interrelationship between technology and subjectivity, tool, and myth, and to pay attention to our students' paying attention. Certainly, technologies are used to manage and "enforce" subjectivities in a multitude of ways. How? To what ends? For whose purposes? More important, how might normalizing uses of technology be resisted, and how might we play our own relationships to technology in ways that expand our sense of available freedom?

In *Non-Discursive Rhetoric: Image and Affect in Multimodal Composition*, Joddy Murray argues that

> this particular time in history is not so much requiring that we apply fundamentally new questions to our pedagogy. Rather, it requires that we revalue and reauthorize what has always been important to our symbol-making process: image and affectivity. We can no longer rest on the assumptions that the body and the mind are separate, that affectivity and "logic" are opposites, or that rhetoric and design are fundamentally separate disciplines. (8)

In a way, we agree with Murray about the significance of rethinking the connection between affectivity and image—indeed, between affectivity, subjectivity, and multimodality in general. We are unsure, however, that such rethinking won't fundamentally change our conception of our pedagogy. Or of our discipline.

Ultimately, we advocate a particular type of recombinant new media literacy, one that is critical and functionally productive, a cyborg synergy that simultanously produces new media (and new media content) and demands critique of the same, and that actively pays attention to how our sense of subjectivity, individually and collectively, changes through our (inter)relationship with technology. Worsham argues that it is necessary to create "a form of critique that links the discursive and the nondiscursive, the working day and the everyday, beyond the mystifying focus on pleasure and desire" (218). In this sense, Haraway still offers us a uniquely powerful way to think about technology and its varied impacts on changing understandings of the (post)human condition. Her "cyborg," as a metaphor marking the slippage of boundaries between human, machine, animal, tool, mythology, gestures toward, even seems to characterize, some of the current complexity of how individuals and groups actually use (and are used by) technologies, particularly the communications technologies that seem nearly ubiquitous on our campuses—Web, tweeting, podcasting, iPods, cell phones, texting, creation of sound and video files, Facebooking, blogging—on a daily basis. To greater and lesser degrees, many of us (students and teachers) use these technologies as a matter of course, creating and consuming content at ever-increasing rates. However, fewer of us use these technologies to reflect consciously on what it means to be human, on what it might mean to be human, on how our sense of humanity, individually and collectively, is potentially enhanced, extended, delimited, estranged, changed, dispersed. Even fewer of us engage in this reflection with our students. Such reflection itself can evoke Haraway's "cyborg myth," since it can serve as an instance of the "transgressed boundaries, potent fusions, and dangerous possibilities which progressive people might explore as one part of needed political work" (154). Indeed, the active, intersubjective use of technology and reflection in the writing classroom could itself be "powerful infidel heteroglossia."

FINAL THOUGHTS

In this final chapter, we have explored an approach to analyzing new media production in the aftermath of a tragedy as a way to model a critical pedagogy that works simultaneously with the content of the tragedy and its mediation. Considering with students how media can be used, purposefully or unconsciously, to discipline and normalize public reactions and discussion is crucial to helping all of us better understand the rhetorical effects—and affects—of multimedia production and circulation. We cannot robustly teach students how to mine the rhetorical affordances of media unless we also enact pedagogies that think about mediation and subjectivity together.

As we consider new and multimedia, we need to recall that Haraway's cyborg myth demands an attention to discursive regimes of the self. It is more than the cyborg identity we might acknowledge at a brute level, with technology insinuating itself into all parts of our lives, from iPods providing personal sound tracks to our manipulation of genetic material. Haraway seeks to carve a space from which to resist those discursive regimes—the drives to normalize, regularize, and control behavior, thought, freedom—since our high technologies offer us, simultaneously, possibilities for greater control and resistance to that control:

> From one perspective, a cyborg world is about the final imposition of a grid of control on the planet, about the final abstraction embodied in a Star Wars apocalypse waged in the name of defence, about the final appropriation of women's bodies in a masculinist orgy of war. From another perspective, a cyborg world might be about lived social and bodily realities in which people are not afraid of their joint kinship with animals and machines, not afraid of permanently partial identities and contradictory standpoints. The political struggle is to see from both perspectives at once because each reveals both dominations and possibilities unimaginable from the other vantage point. (154)

How do we expand our gaze to include multiple perspectives? How might we deploy and even *celebrate* our "permanently partial identities and contradictory standpoints?" We believe a greater attention to the possible histories of new and multimedia is key, for such attention moves us beyond bare functional literacy to the rich multiliteracies made possible by new media. Such attention calls us to move beyond the colonized use of new media to a fuller, cross-disciplinary sense of just what such media offer to our students—and to us.

We return here to "College Saga," the rich student-authored multimedia production that mashed up a *Final Fantasy* narrative and a students' thoughts on college life (Leung). In this text, the overriding and in some ways domineering *Final Fantasy* narrative trope is periodically interrupted by sidebar comments on the Asian American ethnicity of the primary avatar, usually a quirky comment about ethnic and racial assumptions. The "party"—Leung, a Harry Potter look-alike, and a female avatar—encounter an African American Willy Wonka who claims, Star Wars–style, to be Leung's father. Incredulity ensues. While the short scene (one of many such) is played for laughs, it also provides an opportunity to query normative assumptions about who plays such games (not only white boys), as well as what "American" families might look like and be. In other words, within the techno-driven narrative are resistant moments of expansive consciousness. Such moments, we believe, derive from a burgeoning attention to the rhetorical capabilities and subjectivities of new media. They are moments of excess, of composing beyond a genre's expectations in order to resist discursive regimes. How do we bring such moments into the writing classroom?

"Writing," argue Downs and Wardle, is "content—and context—contingent and irreducibly complex" (558). This statement seems particularly true of composing multimodally in new media contexts, which are potentially part of a "powerful infidel heteroglossia." Taken together, irreducibly complex composing and Haraway's infidel heteroglossia suggest an approach to textual production that is excessive, discomposed, and resistant—they suggest ways, that is, to engage the poeticized public sphere that Michael Warner identifies. That's important if our students (and we) are to be critical participants in twenty-first-century literacies. If our field is to more fully engage new and multimedia, that engagement must necessarily take a contextualizing turn, not just a technologizing one. And that contextualizing must concern itself not only with the rhetorical affordances of the technologies we encounter in and out of the classroom—a large task in itself—but also with the discursive regimes of subjectivity and affect that delimit such affordances. Such engagement necessarily pushes against what we might have thought about ourselves as a coherent discipline, for it asks us to work against our own disciplinary legitimacy, if such legitimacy is found in colonizing other fields in service to our own. It asks us to imagine ourselves as "irreducibly complex." It asks us to imagine ourselves as *more.*

NOTES

1. Refiguring Our Relationship to New Media

1. Banks actually writes specifically about technology in his 2006 book, *Race, Rhetoric, and Technology: Searching for Higher Ground.* He imagines technology serving, through archives, older African American rhetorical traditions, but he also wonders how such traditions might be transformed by technology: "How do we bring the communal support of the church play, with parents on the side whispering the lines to chirrens just to get them the experience, or the fact that we're always giving even mediocre speakers applause in a show of support, to digital spaces? What do those practices look like online?" (145).

2. Lev Manovich, in his introductory essay to Wardrip-Fruin and Montfort's collection, notes how even the prospect of defining *new media* requires a historical turn: "Rather than reserving the term 'new media' to refer to the cultural uses of current computer and computer-based network technologies, some authors have suggested that every modern media and telecommunication technology passes through its 'new media stage.' In other words, at some point photography, telephones, cinema, and television each were 'new media.' This perspective redirects our research efforts: rather than trying to identify what is unique about digital computers functioning as media creation, media distribution and telecommunication devices, we may instead look for certain aesthetic techniques and ideological tropes which accompany every new modern media and telecommunication technology at the initial stage of their introduction and dissemination" (19). Indeed, recent major texts and anthologies on new media always historicize; the editors of *New Media: A Critical Introduction*, for example, maintain that "it is a serious flaw not to engage with the histories of the technologies and media under discussion" (Lister, Dovey, Giddings, Grant, and Kelly 4). Particularly

when talking about "new media," the tendency has been to vaunt the "newness" of the media in question, forgetting (purposely or not) the histories that have given rise to the new-fangled. To correct this, anthologists attempt to construct histories of influence and trajectories of technological emergence. Randall Packer and Ken Jordan's edited collection, *Multimedia: From Wagner to Virtual Reality*, contains not only writing by the opera composer and opera house technologist but also work by Italian futurist F. T. Marinetti and avant-garde composer John Cage. And *The New Media Reader*, edited by Noah Wardrip-Fruin and Nick Montfort, offers a rich collection of documents tracing the emergence of contemporary new media, pulling not only from likely sources such as Vannevar Bush, Alan Turing, and Marshall McLuhan, but also from Jorge Luis Borges, William S. Burroughs, and Italo Calvino.

4. Collaboration, Interactivity, and the *Dérive* in Computer Gaming

1. Consider Nicholas Yee's "Understanding MMORPG Addiction" at www.nickyee.com/daedalus/gateway_addiction.html.

2. The work of Hawisher and Selfe may be methodologically useful here, particularly in providing a framework for approaching and understanding how students use new media platforms. In "Becoming Literate in the Information Age: Cultural Ecologies and the Literacies of Technology," Hawisher and Selfe, writing with Brittney Moraski and Melissa Pearson, point out, quite accurately, that "schools are not the sole—and, often, not even the primary—gateways through which people gain access to and practice digital literacies. English composition teachers often have little connection to, and a limited understanding of, the range of literacy practices that happen in digital environments reached through other gateways" (644). To discover more about such literacies, Selfe and Hawisher collected numerous literacy narratives (more than 350) from a variety of individuals willing to be interviewed or fill out questionnaires. Their work, collected and summarized in *Literate Lives in the Information Age: Narratives of Literacy from the United States*, makes a strong case for paying attention to the many divergent and rich literacy practices in our culture. Beyond noting that *access* to technology continues to play a significant role in the extent to which an individual's understanding of literacy is shaped by new media, Selfe and Hawisher point out that "raised and educated in a culture

that valued, and continues to value, alphabetic and print literacies, many . . . teachers remain unsure of how to value new-media literacies, unsure how to practice these new literacies themselves, and unprepared to integrate them at curricular and intellectual levels appropriate for these particular young people" (671).

3. A wide-ranging sampling of such work can be found in anthologies such as *The Video Game Theory Reader*, edited by Mark J. P. Wolf and Bernard Perron. Marc Prensky's *Digital Game-Based Learning* analyzes how games can be used to engage students and promote learning in a variety of contexts, and MIT's Henry Jenkins, professor of humanities, director of MIT's Comparative Media Studies program, and pop culture expert, has created the Education Arcade (www.educationarcade.org/), which studies and assesses the use of computer and video games in educational settings (*STEP*). Conferences and organizations, such as the Games, Learning, and Society Conference (www.glsconference.com) and the Learning Games Initiative (www.mesmernet.org/lgi/), have recently sprung up to study the impact of gaming on learning, as well as the potential of using such games in pedagogical contexts. Academic Gamers (www.academic-gamers.org/) offers a blog to further open up such discussions. Discussions among scholars participating in such groups are often fascinating and complex, focusing primarily on broad social and cultural implications of gaming. Ken S. McAllister's *Game Work: Language, Power, and Computer Game Culture* offers an important critique, ambitiously setting out to develop a "clearer understanding of how the computer game complex has effected individual, communal, and social transformations in the past" (xi).

4. Both students were eager to share information, particularly about the games they play and their history as gamers. There is clearly a generational divide, if not exactly a technological divide, in these students' knowledge about computer technologies. Both students come from fairly middle-class families that had access to computer technologies, though both students reported that their parents were not keen about their sons' gaming. Matt said, "Both of our fathers are very computer savvy. . . . [My dad] played the first Mario game and then after that they started coming out with tons of games, and my father is kind of like a perfectionist so he felt like if he was going to keep playing he would have had to [have] played all

these games, and he believed video games were stupid because it was a waste of time and there were so many to do." Matt's comment about parental disdain is reflected in Mike's comments as well: "My dad absolutely hates video games. . . . He's stomped on some of my video game systems in the past. Like every time I talk to him he says, 'Michael, quit the video.' Which really means quit the video games. He just doesn't know how to say it right for some reason." We imagine that this is a familiar refrain in households comparable to Mike's and Matt's; many parents of this generation may have become increasingly computer savvy but find computer and video gaming relatively useless.

5. Ellen Cushman's comments in "Composing New Media: Cultivating Landscapes of the Mind" may be relevant here. She argues that "new media . . . demands that you perform it, to create it. To do this, it asks you to break from the vending-machine notions of interactivity that are part of what has become conventional design for the World Wide Web" (http://english.ttu.edu/kairos/9.1/binder.html?http://www.msu.edu/%7Ecushmane/one/landscape.html). We sense a creative interactivity in Mike's and Matt's engagement with their MMORPGs—one that embrace multitasking, multimodality, and multiple goals for play.

6. As such, we believe he might enjoy Ken McAllister's recent work on gaming and labor, *Game Work: Language, Power, and Computer Game Culture.*

7. In both books cited in our study, Kress has much to say about the multimodal nature of contemporary textuality, but his accounting seems curiously ahistorical, or historical only in the sense that it seems to turn on a textually driven past now giving way to multimodal future. Such turns are hardly as dramatic as we tend to think they are, and we pause to ask, as in previous chapters, where is the long history of the avant-garde? Kress offers no accounting of it, yet as we have seen, the avant-garde often mixes and matches textuality and visuality, reimagining them in startling ways.

8. Zoevera Ann Jackson is interested in connecting video/computer gaming practices to the teaching of narrative, and she provides a number of exercises and activities to guide instructors interested

in making such connections for their students. For instance, one exercise asks students to create and describe their own video games.

9. Scholarly attention to issues of gaming and gender has tended to focus on women's (lack of) access to, use of, and experience with computer games. See early work by Don Tapscott in *Growing Up Digital* and Helen Cunningham in "Moral Kombat and Computer Game Girls." Thurlow, Lengel, and Tomic summarize research about women and gaming and argue that, while computer games are largely played by men, there is a growing minority of women players in the United States; further, they suggest that "women are more likely to participate in online chat and emailing than men." They also note, however, that "feminist scholars argue that a majority of female stereotypes in gaming are also associated with violence—and these characters are created for men to play" (131). See also more recent work: Karen Orr Vered's "Blue Group Boys Play *Incredible Machine,* Girls Play Hopscotch: Social Discourse and Gendered Play at the Computer"; Justine Cassell and Henry Jenkins's edited collection, *From Barbie to Mortal Kombat: Gender and Computer Games*; Sheri Graner Ray's *Gender Inclusive Game Design: Expanding the Market*; and Lori Kendall's recent *Hanging Out in the Virtual Pub: Masculinities and Relationships Online,* which examines hegemonic masculinities as they are performed, maintained, and challenged in some online spaces, including some gamespaces.

5. Theorizing the Multimodal Subject

1. Much early work in cyberculture studies has focused on issues of subjectivity, such as the creation and dissemination of online identities. Sherry Turkle's pioneering work in *Life on the Screen*, for instance, drew our attention to the possibilities for exploring identities (and communities) in online spaces, and no doubt part of the draw for the millions of users of Second Life and many MMORPGs is the chance to proliferate and experiment with multiple identities.

2. It is a sad reality that neither the Virginia Tech tragedy nor the human response to it is unique. Cell phones, texting, and amateur video have played a role in every major disaster since the technologies became readily available. Dylan Klebold and Eric Harris, for example, documented their plans for Columbine High School on videotapes, a number of which were found in Harris's bedroom after the massacre, and there are, literally, terabytes of digital ar-

chiving and commentary on 9/11, Hurricane Katrina, the 2004 tsunami in Southeast Asia, the 2005 London subway bombings, and roadside ambushes in Iraq and Afghanistan. Our profession and others have responded to trauma and its implications for our work: witness Shane Borrowman's 2005 collection, *Trauma and the Teaching of Writing*; the 2004 two-volume issue of *JAC: A Journal of Rhetoric, Culture, and Politics* focused on "trauma and rhetoric"; online discussions on the WPA listserv about using writing and the composition class to respond to institution-wide tragedies; and, of course, the burgeoning field of trauma studies. Indeed, the sad, simultaneous proliferation of technology and tragedy has offered much evidence of the epistemological power of writing; to write is to make sense, even if what we write about is, finally, senseless.

3. See Gibbs and Roche's "The Columbine Tapes" at *Time.com* for more information about the Columbine shooting and the shooters' use of video and other technology.

4. Dissenting views on the blog site appeared scattered throughout the postings:

> The thing is, who would you report him to? What would you say? You can't lock someone up for being "strange". If so many, many people would be in jail and then there's the question of civil liberties. So until the tragedy happens, there's just nothing you can do.
>
> You can't 'turn someone in' because they create a piece of art that you think may say something about them. It's a work of imagination (or it's supposed to be). Perhaps he was just getting inside the mind of a young sociopath. Turns out he had issues obviously, but you know how many thousands of plays, stories and poems are submitted to creative writing classes that are way more disturbing than that? You can't just assume that means someone is demented.
>
> It wasn't that deranged. I have seen way more deranged in films. ("Cho")

5. Like many of our colleagues in English and writing studies across the country, we sympathized with our colleagues at Virginia Tech and understood that writing and literature courses would be among

the primary places—given their size and the humanist content and subjects frequently taught in them—in which students (and faculty) would want to process such a terrifying and tragic experience. We also understood that Cho's status as an English major, and the fact that both his print and video texts were held up as objects for scrutiny and even as "explanations" for his behavior, demanded an accounting of the connections between violence, writing, and subjectivity. We know we are not alone in our continuing horror in response to that April morning in Virginia. We wonder, again, how we as a culture might prevent such violence, and we are keenly aware of the fundamental inability of academic texts to respond to such a tragedy. We therefore offer this chapter as an exploration of yet another explosive instance of what Lynn Worsham famously called "pedagogic violence." Indeed, such tragedies as the Virginia Tech murders pose seemingly unanswerable questions: Why would someone do such a thing? What kind of person is capable of killing so many others? What must his sense of self, his interior life, have been like? And how have his actions changed the interior and communal lives of others? Such questions cut to the heart of subjectivity, and they were frequently debated through a wide variety of electronic media.

6. Some of our previous work has touched on this idea; specifically, see Jonathan's *Digital Youth: Emerging Literacies on the World Wide Web*, which examines students' development of rhetorical savvy in the design of websites for a variety of purposes—personal, communal, and even political.

WORKS CITED

Adler-Kassner, Linda, John Majewski, and Damian Koshnick. "The Value of Troublesome Knowledge." *Composition Forum* 26 (2012): n. pag. Web. 1 Sept. 2012.

Alexander, Jonathan. "Digital Spins: The Pedagogy and Politics of Student-Centered E-Zines." *Computers and Composition* 19.4 (2002): 387–410. Print.

———. *Digital Youth: Emerging Literacies on the World Wide Web*. Creskill, NJ: Hampton, 2006. Print.

Alexander, Jonathan, and Elizabeth Losh. "Whose Literacy Is It Anyway? Examining a First-Year Approach to Gaming across Curricula." *Currents in Electronic Literacy* 13 (2010): n. pag. Web. 15 Feb. 2012.

Alexander, Jonathan, and Jacqueline Rhodes. "Queerness: An Impossible Subject for Composition." *JAC: A Journal of Rhetoric, Culture, and Politics* 31.1–2 (2011): 711–40. Print.

———. "Queerness, Multimodality, and the Possibilities of Re/Orientation." *Composing (Media) = Composing (Embodiment): Bodies, Technologies, Writing, the Teaching of Writing*. Ed. Kristin L. Arola and Anne Frances Wysocki. Logan: Utah State UP, 2012. 187–211. Print.

Anderson, Daniel. "Prosumer Approaches to New Media Composition: Consumption and Production in Continuum." *Kairos* 8.1 (2003): n. pag. Web. 15 Feb. 2012.

apesmen09. "Multimodal Literacy Project." *YouTube*. YouTube, 4 Nov. 2009. Web. 9 Aug. 2013.

"Assignments." *WR39B & WR 37: Critical Reading and Rhetoric*. Dec. 2012. Web. 11 Aug. 2013.

Baca, Damián. *Mestiz@ Scripts, Digital Migrations, and the Territories of Writing*. New York: Palgrave Macmillan, 2008. Print.

Bailie, Brian. "'If You Don't Believe That You're Doing Some Good with the Work That You Do, Then You Shouldn't Be Doing It': An Interview with Cindy Selfe." *Composition Forum* 21 (2010): n. pag. Web. 15 Feb. 2012.

Bakardjieva, Maria. *Internet Society: The Internet in Everyday Life.* Thousand Oaks: Sage, 2005. Print.

Ball, Cheryl E., and James Kalmbach, eds. *RAW: (Reading and Writing) New Media.* Cresskill: Hampton, 2010. Print.

Banks, Adam J. *Digital Griots: African American Rhetoric in a Multimedia Age.* Carbondale: Southern Illinois UP, 2011. Print.

———. *Race, Rhetoric, and Technology: Searching for Higher Ground.* Mahwah: Erlbaum, 2006. Print.

Baron, Dennis. *A Better Pencil: Readers, Writers, and the Digital Revolution.* New York: Oxford UP, 2009. Print.

———. "From Pencils to Pixels: The Stages of Literacy Technologies." Hawisher and Selfe, *Passions* 15–33.

Barron, Monica. "Creative Writing Class as Crucible." *Academe* 93.6 (2007): 40–43. Print.

Bartholomae, David. "Inventing the University." *Journal of Basic Writing* 5 (1986): 4–23. Print.

Bawarshi, Anis. *Genre and the Invention of the Writer: Reconsidering the Place of Invention in Composition.* Logan: Utah State UP, 2003. Print.

Beard, Jeannie Parker. "Documenting Arguments, Proposing Change: Reflections on Student-Produced Proposal Documentaries." *Computers and Composition Online* (Spring 2010): n. pag. Web. 17 Nov. 2013.

beautifulataxia2. "Digital Literacy Narrative." *YouTube.* YouTube, 18 Nov. 2008. Web. 9 Aug. 2013.

Beauty and the Beast. Dir. Gary Trousdale and Kirk Wise. Disney, 1991. DVD.

La Belle et la Bête. Dir. Jean Cocteau. 1946. Criterion Collection, 2003. DVD.

"Bio of Kkir." *Thrall's Chosen of Dalaran. Guildportal.com.* Web. 17 Nov. 2013.

Birkerts, Sven. *The Gutenberg Elegies: The Fate of Reading in an Electronic Age.* London: Faber and Faber, 2006. Print.

Blakesley, David, and Collin Brooke. "Introduction: Notes on Visual Rhetoric." *Enculturation* 3.2 (2001): n. pag. Web. 9 Feb. 2012.

Bloom, Lynn Z., Donald A. Daiker, and Edward M. White, eds. *Composition in the Twenty-First Century: Crisis and Change.* Carbondale: Southern Illinois UP, 1996. Print.

———. *Composition Studies in the New Millennium: Rereading the Past, Rewriting the Future.* Carbondale: Southern Illinois UP, 2003. Print.

Bolter, Jay David, and Richard Grusin. *Remediation: Understanding New Media.* Cambridge: MIT P, 1999. Print.

Borrowman, Shane, ed. *Trauma and the Teaching of Writing.* Albany: State U of New York P, 2005. Print.

Bowen, Tracey, and Carl Whithaus, eds. *Multimodal Literacies and Emerging Genres.* Pittsburgh: U of Pittsburgh P, 2013. Print.

Brandt, Deborah. *Literacy in American Lives.* New York: Oxford UP, 2001. Print.

Brin, David. *The Transparent Society: Will Technology Force Us to Choose between Privacy and Freedom?* Reading: Addison-Wesley, 1998. Print.

Brooke, Collin Gifford. *Lingua Fracta: Towards a Rhetoric of New Media.* Cresskill: Hampton, 2009. Print.

Bulletin of Courses 2012–2014 for California State University San Bernardino. San Bernardino: California State U, 2012. Print.

Cassell, Justine, and Henry Jenkins, eds. *From Barbie to Mortal Kombat: Gender and Computer Games.* Cambridge: MIT P, 1998. Print.

Chang, Edmond Y. "Gaming as Writing, or, *World of Warcraft* as World of Wordcraft." *Computers and Composition Online* (Fall 2008): n. pag. Web. 15 Feb. 2012.

ChoLovingVampire. "The Warrior Seung Hui Cho." *YouTube.* YouTube, 23 May 2010. Web. 1 Sept. 2012.

"Cho Seung-Hui's Plays." *News Bloggers.* 17 Apr. 2007. Web. 31 July 2009.

Colby, Rebekah Shultz, and Richard Colby. "A Pedagogy of Play: Integrating Computer Games into the Writing Classroom." *Computers and Composition* 25.3 (2008): 300–12. Print.

Conference on College Composition and Communication (CCCC). "CCCC Position Statement on Teaching, Learning, and Assessing Writing in Digital Environments." 25 Feb. 2004. Web. 9 Feb. 2012.

———. "CCCC Statement on the Multiple Uses of Writing." 19 Nov. 2007. Web. 8 Aug. 2013.

———. "Principles and Practices in Electronic Portfolios." 19 Nov. 2007. Web. 9 Feb. 2012.

Council of Writing Program Administrators. "WPA Outcomes Statement for First-Year Composition." *WPAcouncil.org.* 2000. Web. 9 Feb. 2012.

"Creative Writing and the Virginia Tech Massacre." *Chronicle of Higher Education.* 11 May 2007. Web. 13 Aug. 2013.

Cunningham, Helen. "Moral Kombat and Computer Game Girls." *In Front of the Children: Screen Entertainment and Young Audiences.* Ed. Cary Bazalgette and David Buckingham. London: British Film Institute, 1995. 188–200. Print.

Cushman, Ellen. "Composing New Media: Cultivating Landscapes of the Mind." *Kairos* 9.1 (2004): n. pag. Web. 9 Feb. 2012.

Debord, Guy. *The Society of the Spectacle.* Trans. Donald Nicholson-Smith. New York: Zone, 1995. Print. Trans. of *La Société du Spectacle.* Paris: Buchet/Chastel, 1967.

de Certeau, Michel. *The Practice of Everyday Life.* Trans. Steven F. Rendall. Berkeley: U of California P, 1984. Print.

"Definitions." *Situationist International Online.* Web. 11 Aug. 2013.

DelBarrio, LaReina. "Pastor Rick Warren—Saddleback Homophobia Child Molester Church California Religious Bigotry!" *YouTube.* YouTube. 12 Dec. 2008. Web. 17 Nov. 2013.

Devitt, Amy J., Anis Bawarshi, and Mary Jo Reiff. "Materiality and Genre in the Study of Discourse Communities." *College English* 65.5 (2003): 541–58. Print.

DeVoss, Dànielle Nicole, Joseph Johansen, Cynthia L. Selfe, and John C. Williams, Jr. "Under the Radar of Composition Programs: Glimpsing the Future through Case Studies of Literacy in Electronic Contexts." Bloom, Daiker, and White, *Composition Studies in the New Millennium* 157–73.

DeVoss, Dànielle Nicole, and Julie Platt. "Image Manipulation and Ethics in a Digital-Visual World." *Computers and Composition Online* (Fall 2011): n. pag. Web. 9 Feb. 2012.

Disturbed. "Indestructible." *Indestructible.* Reprise, 2008. CD.

Dobrin, Sidney I. "Through Green Eyes: Complex Visual Culture and Post-Literacy." *Environmental Education Research* 16.3–4 (2010): 265–78. Print.

Downs, Douglas, and Elizabeth Wardle. "Teaching about Writing, Righting Misconceptions: (Re)Envisioning 'First-Year Composition' as 'Introduction to Writing Studies.'" *College Composition and Communication* 58.4 (2007): 552–84. Print.

Dubisar, Abby M., and Jason Palmeri. "Palin/Pathos/Peter Griffin: Political Video Remix and Composition Pedagogy." *Computers and Composition* 27.2 (2010): 77–93. Print.

Eason, Kat. "Sample Blog Prompt." *39b-Staff listserv.* 19 Jan. 2010. Web. 11 Aug. 2013.

———. "Workshop Follow-up: Close-Read a Scene Project." *39b-Staff listserv.* 25 Jan. 2010. Web. 11 Aug. 2013.

Faigley, Lester. "The Challenge of the Multimedia Essay." Bloom, Daiker, and White 174–87.

Foucault, Michel. *Discipline and Punish: The Birth of the Prison.* Trans. Alan Sheridan. New York: Random, 1977. Print.

———. *Ethics: Subjectivity and Truth.* Ed. Paul Rabinow. Trans. Robert Hurley. New York: New, 1997. Print. Volume 1 of *Essential Works of*

Foucault, 1954–1984.

———. *The History of Sexuality, Volume 1: An Introduction.* Trans. Robert Hurley. New York: Vintage, 1980. Print.

Frasca, Gonzalo. "Videogames of the Oppressed: Critical Thinking, Education, Tolerance, and Other Trivial Issues." *First Person: New Media as Story, Performance, and Game.* Ed. Noah Wardrip-Fruin and Pat Harrigan. Cambridge: MIT P, 2004. 85–94. Print.

Fregoso, Rosa Linda. *MeXicana Encounters: The Making of Social Identities on the Borderlands.* Berkeley: U of California P, 2003. Print.

Fuentes, Carlos. "The Discreet Charm of Luis Bunuel." *New York Times Magazine* 11 Mar. 1973: 93. Print.

Garber, Linda. *Identity Poetics: Race, Class, and the Lesbian-Feminist Roots of Queer Theory.* New York: Columbia UP, 2001. Print.

Gee, James Paul. *What Video Games Have to Teach Us about Learning and Literacy.* Rev. and updated ed. New York: Palgrave Macmillan, 2007. Print.

Gee, James Paul, and Elisabeth R. Hayes. *Language and Learning in the Digital Age.* New York: Routledge, 2011. Print.

Gibbs, Nancy, and Timothy Roche. "The Columbine Tapes." *Time.com.* Time, 20 Dec. 1999. Web. 11 Aug. 2013.

Girard, René. *Deceit, Desire, and the Novel: Self and Other in Literary Structure.* Trans. Yvonne Freccero. Baltimore: Johns Hopkins UP, 1976. Print.

Gogan, Brian J. "Digital Literacy Narrative Assignment." Personal website. *Briangogan.com.* Web. 11 Aug. 2013.

Goggin, Peter N., and Maureen Daly Goggin. "Presence in Absence: Discourses and Teaching (in, on, and about) Trauma." Borrowman 29–51.

Gould, Sarah. "A Closer Look into Physical Disabilities: An Oral History Video." *TheJUMP* 2.1 (2010). Web. 9 Feb. 2012.

Gray-Rosendale, Laura, and Sibylle Gruber, eds. *Alternative Rhetorics: Challenges to the Rhetorical Tradition.* Albany: State U of New York P, 2001. Print.

Groktal. "Guide Writing." *WoW-Pro.com.* Web. 11 Aug. 2013.

Handa, Carolyn, ed. *Visual Rhetoric in a Digital World: A Critical Sourcebook.* Boston: Bedford/St. Martin's, 2004. Print.

Haraway, Donna J. *Simians, Cyborgs, and Women: The Reinvention of Nature.* New York: Routledge, 1991. Print.

Hawisher, Gail E., and Cynthia L. Selfe, eds. *Global Literacies and the World-Wide Web.* New York: Routledge, 2000. Print.

———. *Passions, Pedagogies, and 21st Century Technologies.* Logan: Utah State UP, 1999. Print.

Hawisher, Gail E., and Cynthia L. Selfe (with Brittney Moraski and Melissa Pearson). "Becoming Literate in the Information Age: Cultural Ecologies and the Literacies of Technology." *College Composition and Communication* 55.4 (2004): 642–92. Print.

Hawk, Byron. *A Counter-History of Composition: Toward Methodologies of Complexity*. Pittsburgh: U of Pittsburgh P, 2007. Print.

Henthorne, Tom. "Cyber-Utopias: The Politics and Ideology of Computer Games." *Studies in Popular Culture* 25.3 (2003): 63–76. Print.

Hesse, Doug. "Response to Cynthia L. Selfe's 'The Movement of Air, the Breath of Meaning: Aurality and Multimodal Composing.'" *College Composition and Communication* 61.3 (2010): 602–05. Print.

———. "Saving a Place for Essayistic Literacy." Hawisher and Selfe, *Passions* 34–48.

Hocks, Mary E. "Understanding Visual Rhetoric in Digital Writing Environments." *College Composition and Communication* 54.4 (2003): 629–56. Print.

Hocks, Mary E., and Michelle R. Kendrick, eds. *Eloquent Images: Word and Image in the Age of New Media*. Cambridge: MIT P, 2003. Print.

Hogan, Monika I. "'Still Me on the Inside, Trapped': Embodied Captivity and Ethical Narrative in Leslie Feinberg's *Stone Butch Blues*." *Third-Space: A Journal of Feminist Theory and Culture* 3.2 (2004): n. pag. Web. 11 Aug. 2013.

Jackson, Zoevera Ann. "Connecting Video Games and Storytelling to Teach Narratives in First-Year Composition." *Kairos* 7.3 (Fall 2002): n. pag. Web. 15 Feb. 2012.

Jenkins, Henry. *Fans, Bloggers, and Gamers: Media Consumers in a Digital Age*. New York: New York UP, 2006. Print.

Journal for Undergraduate Multimedia Projects (TheJUMP), The. About page. *The JUMP*. The University of Texas at Austin. Web. 30 Oct. 2013.

Kendall, Lori. *Hanging Out in the Virtual Pub: Masculinities and Relationships Online*. Berkeley: U of California P, 2002. Print.

Kim, Kyle. "Closer." *TheJUMP* 2.1 (2010). Web. 17 Nov. 2013.

King, Brad, and John Borland. *Dungeons and Dreamers: The Rise of Computer Game Culture from Geek to Chic*. New York: McGraw, 2003. Print.

Kinney, Kelly, Thomas Girshin, and Barrett Bowlin. "The Third Turn Toward the Social: Nancy Welch's *Living Room*, Tony Scott's *Dangerous Writing*, and Rhetoric and Composition's Turn toward Grassroots Political Activism." *Composition Forum* 21 (2010): n. pag. Web. 9 Feb. 2012.

Kopp, Drew, and Sharon McKenzie Stevens. "Re-Articulating the Mission and Work of the Writing Program with Digital Video." *Kairos* 15.1 (2010): n. pag. Web. 9 Feb. 2012.

Kress, Gunther. *Literacy in the New Media Age*. New York: Routledge, 2003. Print.

———. *Multimodality: A Social Semiotic Approach to Contemporary Communication*. New York: Routledge, 2010. Print.

Laguna Art Museum. "*WoW*: Emergent Media Phenomenon." *lagunaartmuseum.org*. 2009. Web. 17 Nov. 2013.

Lambourn, Ryan. "V-Tech Rampage." *Newgrounds.com*. Web. 9 Aug. 2013.

Lanham, Richard A. *The Economics of Attention: Style and Substance in the Age of Information*. Chicago: U of Chicago P, 2006. Print.

Lauer, Janice M. "Rhetoric and Composition." McComiskey 106–52.

Leung, Mark. "College Saga." *YouTube*. YouTube, 24 Dec. 2006. Web. 9 Aug. 2013.

Lieberman, Max. "Four Ways to Teach with Video Games." *Currents in Electronic Literacy* 2010: n. pag. Web. 15 Feb. 2012.

Linder, Fletcher. "Speaking of Bodies, Pleasures, and Paradise Lost: Erotic Agency and Situationist Ethnography." *Cultural Studies* 15.2 (2001): 352–74. Print.

Lindsay F. "Literacy Narrative Project." *YouTube*. YouTube, 5 Feb. 2010. Web. 17 Nov. 2013.

Lister, Martin, Jon Dovey, Seth Giddings, Iain Grant, and Kieran Kelly, eds. *New Media: A Critical Introduction*. 2nd ed. New York: Routledge, 2009. Print.

Losh, Elizabeth. "Digital Rhetoric: Becoming a Conscious and Critical User of Social Media." Web. 11 Aug. 2013.

Lutkewitte, Claire, ed. *Multimodal Composition: A Critical Sourcebook*. Boston: Bedford/St. Martin's, 2013. Print.

———. "Web 2.0 Technologies in First-Year Writing." *Computers and Composition Online* (Spring 2009): n. pag. Web. 9 Feb. 2012.

Lyotard, Jean-François, and Jean Loup Thebaud. *Just Gaming*. Trans. Wlad Godzich. Minneapolis: U of Minnesota P, 1985. Print.

Mailloux, Steven. *Disciplinary Identities: Rhetorical Paths of English, Speech, and Composition*. New York: MLA, 2006. Print.

Manovich, Lev. *The Language of New Media*. Cambridge: MIT P, 2001. Print.

———. "New Media from Borges to HTML." Wardrip-Fruin and Montfort 13–25.

Mansfield, Nick. *Subjectivity: Theories of the Self from Freud to Haraway*. New York: New York UP, 2000. Print.

Mao, LuMing. *Reading Chinese Fortune Cookie: The Making of Chinese American Rhetoric*. Logan: Utah State UP, 2006. Print.

Marx, Karl. *Capital: A Critique of Political Economy*. Vol. 1. Trans. Ben Fowkes. New York: Penguin, 1990. Print.

McAllister, Ken S. *Game Work: Language, Power, and Computer Game Culture*. Tuscaloosa: U of Alabama P, 2004. Print.

McComiskey, Bruce, ed. *English Studies: An Introduction to the Discipline(s)*. Urbana: NCTE, 2006. Print.

McCorkle, Ben. *Rhetorical Delivery as Technological Discourse: A Cross-Historical Study*. Carbondale: Southern Illinois UP, 2012. Print.

McIver, Gillian. "Media and the Spectacular Society." *The Hypermedia Research Centre*. 1997. Web. 15 Feb. 2012.

Meeks, Melissa, and Alex Ilyasova. "A Review of Digital Video Production in Post-Secondary English Classrooms at Three Universities." *Kairos* 8.2 (2003): n. pag. Web. 9 Feb. 2012.

Micciche, Laura. "Emotion, Ethics, and Rhetorical Action." *JAC* 25.1 (2005): 161–84. Print.

Miller, Richard E. "The Fear Factor." *Academe* 93.6 (2007): 33–37. Print.

Miller, Susan, ed. *The Norton Book of Composition Studies*. New York: Norton, 2009. Print.

———. *Trust in Texts: A Different History of Rhetoric*. Carbondale: Southern Illinois UP, 2008. Print.

———. "Why Composition Studies Disappeared and What Happened Then." Bloom, Daiker, and White, *Composition Studies in the New Millennium* 48–56.

Moran, Charles. "*Computers and Composition* 1983–2002: What We Have Hoped For." *Computers and Composition* 20 (2003): 343–58. Print.

Mulholland Drive. Dir. David Lynch. 2001. Universal Studios, 2002. DVD.

Murray, Janet H. *Hamlet on the Holodeck: The Future of Narrative in Cyberspace*. Boston: MIT P, 1998. Print.

Murray, Joddy. *Non-Discursive Rhetoric: Image and Affect in Multimodal Composition*. Albany: State U of New York P, 2009. Print.

Nardi, Bonnie A. *My Life as a Night Elf Priest: An Anthropological Account of* World of Warcraft. Ann Arbor: U of Michigan P, 2010. Print.

Neal, Michael R. "Assessment in the Service of Learning." *College Composition and Communication* 61.4. (2010): 746–58. Print.

newartisticdirection. "I.D. / self :: the new 'real.'" *YouTube*. YouTube, 7 Dec. 2008. Web. 9 Aug. 2013.

Newcomb, Matthew J. "Taking Attendance: Absent Writing and the Value of Suffering." *JAC* 24.3 (2004): 755–84. Print.

Packer, Randall, and Ken Jordan, eds. *Multimedia: From Wagner to Virtual Reality*. New York: Norton, 2001. Print.

Palmeri, Jason. *Remixing Composition: A History of Multimodal Writing Pedagogy.* Carbondale: Southern Illinois UP, 2012. Print.

"The Paragonian Knights." *Guildportal.com*. Web. 11 Aug. 2013.

Paul, Christopher. "World of Rhetcraft: Rhetorical Production and Raiding in *World of Warcraft*." *Writing and the Digital Generation: Essays on New Media Rhetoric*. Ed. Heather Urbanski. Jefferson: McFarland, 2010. 152–61. Print.

Penrod, Diane. *Composition in Convergence: The Impact of New Media on Writing Assessment.* Mahway: Erlbaum, 2005. Print.

Porter, James E. "Recovering Delivery for Digital Rhetoric." *Computers and Composition* 26.4 (2009): 207–24. Print.

Postman, Neil. *Technopoly: The Surrender of Culture to Technology.* New York: Vintage, 1993. Print.

Prensky, Marc. *Digital Game-Based Learning.* New York: McGraw, 2001. Print.

Rand, Erica. "Doing It in Class: On the Payoffs and Perils of Teaching Sexually Explicit Queer Images." *Radical Teacher* 45 (1994): 29–32. Print.

Ray, Sheri Graner. *Gender Inclusive Game Design: Expanding the Market.* Hingham: Charles River Media, 2004. Print.

Reiff, Mary Jo, and Anis Bawarshi. "Tracing Discursive Resources: How Students Use Prior Genre Knowledge to Negotiate New Writing Contexts in First-Year Composition." *Written Communication* 28.3 (2011): 312–37. Print.

Rivers, Nathaniel A., and Ryan P. Weber. "Ecological, Pedagogical, Public Rhetoric." *College Composition and Communication* 63.2 (2011): 187–218. Print.

Rodriguez, Richard. "Digital Literacy Narrative final_0001.avi." *YouTube.* YouTube, 22 Feb. 2010. Web. 29 Nov. 2013.

Sánchez, Raúl. *The Function of Theory in Composition Studies*. Albany: State U of New York P, 2005. Print.

Schroeder, Christopher. *ReInventing the University: Literacies and Legitimacy in the Postmodern Academy.* Logan: Utah State UP, 2001. Print.

Schroeder, Christopher, Helen Fox, and Patricia Bizzell, eds. *ALT DIS: Alternative Discourses and the Academy*. Portsmouth: Heinemann-Boynton/Cook, 2002. Print.

Sedgwick, Eve Kosofsky. *Tendencies*. Durham: Duke UP, 1993. Print.

Selber, Stuart A. *Multiliteracies for a Digital Age*. Carbondale: Southern Illinois UP, 2004. Print.

———, ed. *Rhetoric and Technologies: New Directions in Writing and Communication*. Columbia: U of South Carolina P, 2010. Print.

Selfe, Cynthia L. "The Movement of Air, the Breath of Meaning: Aurality and Multimodal Composing." *College Composition and Communication* 60.4 (2009): 616–63. Print.

———. "Students Who Teach Us: A Case Study of a New Media Text Designer." Wysocki et al. 43–66.

———. *Technology and Literacy in the 21st Century: The Importance of Paying Attention*. Carbondale: Southern Illinois UP, 1999. Print.

———. "Toward New Media Texts: Taking Up the Challenge of Visual Literacy." Wysocki et al. 67–110.

Selfe, Cynthia L., and Gail E. Hawisher, eds. *Gaming Lives in the Twenty-First Century: Literate Connections*. New York: Palgrave Macmillan, 2007. Print.

———. *Literate Lives in the Information Age: Narratives of Literacy from the United States*. Mahwah: Erlbaum, 2004. Print.

Selfe, Cynthia L., Anne F. Mareck, and Josh Gardiner. "Computer Gaming as Literacy." Selfe and Hawisher, *Gaming* 21–35.

Selfe, Richard J., and Cynthia L. Selfe. "Critical Technological Literacy and English Studies: Teaching, Learning, and Action." Yagelski and Leonard 344–81.

SeungHuiChoHero. "About." *YouTube*. YouTube, 24 Nov. 2010. Web. 13 Aug. 2013.

Sheridan, David M., Jim Ridolfo, and Anthony J. Michel. "The Available Means of Persuasion: Mapping a Theory and Pedagogy of Multimodal Public Rhetoric." *JAC* 25.4 (2005): 803–44. Print.

———. *The Available Means of Persuasion: Mapping a Theory and Pedagogy of Multimodal Public Rhetoric*. Anderson: Parlor, 2012. Print.

Shipka, Jody. *Toward a Composition Made Whole*. Pittsburgh: U of Pittsburgh P, 2011. Print.

Sidler, Michelle, Richard Morris, and Elizabeth Overman Smith, eds. *Computers in the Composition Classroom: A Critical Sourcebook*. Boston: Bedford/St. Martin's, 2008. Print.

"The Silvermoon Sisterhood: A World of Warcraft RPPvP Guild." *SilvermoonSisterhood.com*. Web. 15 Feb. 2012.

Sirc, Geoffrey. "Box-Logic." Wysocki et al. 111–46.

———. *English Composition as a Happening*. Logan: Utah State UP, 2002. Print.

———. "'What Is Composition . . . ?' After Duchamp (Notes toward a General Teleintertext)." Hawisher and Selfe, *Passions* 178–204.

Socolovsky, Maya. "Cyber-Spaces of Grief: Online Memorials and the Columbine High School Shootings." *JAC* 24.2 (2004): 467–89. Print.

STEP: The Education Arcade. Massachusetts Institute of Technology, Scheller Teacher Education Program. Web. 31 Oct. 2013.

Stracey, Frances. "Surviving History: A Situationist Archive." *Art History* 26.1 (2003): 56–77. Print.

Tapscott, Don. *Growing Up Digital: The Rise of the Net Generation.* New York: McGraw, 1998. Print.

ThePelly. "KEKEKE—The Ballad of Cho Seung-Hui." *YouTube.* YouTube, 24 Aug. 2007. Web. 9 Aug. 2013.

Therborn, Gören. *The Ideology of Power and the Power of Ideology.* London: Verso, 1980. Print.

"The Third Faction: Missions from the *World of Warcraft* Art Corps." *Thirdfaction.org.* Web. 11 Aug. 2013.

39BUCI. "A Costly Increase." *YouTube.* YouTube, 4 Dec. 2009. Web. 11 Aug. 2013.

"Thrall's Chosen." *Guildportal.com.* Web. 11 Aug. 2013.

Thurlow, Crispin, Laura Lengel, and Alice Tomic. *Computer Mediated Communication: Social Interactions and the Internet.* London: Sage, 2004. Print.

Toogie. "A Newcomer's Guide to Warcraft—Toogie Learned the Hard Way So You Don't Have To!" *Wow-Pro.com.* Web. 11 Aug. 2013.

Turkle, Sherry. *Life on the Screen: Identity in the Age of the Internet.* New York: Simon, 1995. Print.

Vaneigem, Raoul. *The Revolution of Everyday Life.* Trans. Donald Nicholson-Smith. London: Aldgate, 1994. Print.

Vered, Karen Orr. "Blue Group Boys Play *Incredible Machine*, Girls Play Hopscotch: Social Discourse and Gendered Play at the Computer." *Digital Diversions: Youth Culture in the Age of Multimedia.* Ed. Julian Sefton-Green. London: UCL P, 1998. 43–61. Print.

Vielstimmig, Myka. "Petals on a Wet, Black Bough: Textuality, Collaboration, and the New Essay." Hawisher and Selfe, *Passions* 89–114.

Virilio, Paul. *The Information Bomb.* Trans. Chris Turner. London: Verso, 2000. Print.

Wallace, David L., and Jonathan Alexander. "Queer Rhetorical Agency: Questioning Narratives of Heteronormativity." *JAC* 29.4 (2009): 793–819. Print.

Wardrip-Fruin, Noah, and Nick Montfort, eds. *The New Media Reader.* Cambridge: MIT P, 2003. Print.

Wark, McKenzie. *Gamer Theory*. Cambridge: Harvard UP, 2007. Print.

Warner, Michael. *Publics and Counterpublics*. Brooklyn: Zone, 2002. Print.

Weisser, Christian R. *Moving Beyond Academic Discourse: Composition Studies and the Public Sphere*. Carbondale: Southern Illinois UP, 2002. Print.

Welch, Nancy. *Living Room: Teaching Public Writing in a Privatized World*. Portsmouth: Boynton/Cook, 2008. Print.

Westbrook, Steve. "Visual Rhetoric in a Culture of Fear: Impediments to Multimedia Production." *College English* 68.5 (2006): 457–80. Print.

Wolf, Mark J. P., and Bernard Perron, eds. *The Video Game Theory Reader*. London: Routledge, 2003. Print.

Wolff, Bill. "Remixing Composition in the Writing Classroom: An Installation of Student Videos." Online installation and presentation at the Computers and Writing Online Conference. West Lafayette, IN. May 2010. Web. 9 Feb. 2012.

Wooten, Eric. "Literacy Narrative." *YouTube*. YouTube, 18 Sept. 2008. Web. 9 Aug. 2013.

World of Warcraft Pro. Web. 11 Aug. 2013.

Worsham, Lynn. "Going Postal: Pedagogic Violence and the Schooling of Emotion." *JAC* 18.2 (1998): 213–45. Print.

"Writing Requirements." *UC Irvine Campus Writing Coordinator*. Web. 11 Aug. 2013.

Wysocki, Anne Frances. "awaywithwords: On the Possibilities in Unavailable Designs." *Computers and Composition* 22.1 (2005): 55–62. Print.

Wysocki, Anne Frances, and Johndan Johnson-Eilola. "Blinded by the Letter: Why Are We Using Literacy as a Metaphor for Everything Else?" Hawisher and Selfe, *Passions* 349–68.

Wysocki, Anne Frances, Johndan Johnson-Eilola, Cynthia L. Selfe, and Geoffrey Sirc. *Writing New Media: Theory and Applications for Expanding the Teaching of Composition*. Logan: Utah State UP, 2004. Print.

Yagelski, Robert P., and Scott A. Leonard, eds. *The Relevance of English: Teaching That Matters in Students' Lives*. Urbana: NCTE, 2002. Print.

Yancey, Kathleen Blake. "Made Not Only in Words: Composition in a New Key." *College Composition and Communication* 56.2 (2004): 297–328. Print.

Yee, Nicholas. "MMORPG Addiction." *The Daedalus Gateway: The Psychology of MMORPGs*. 2004. Web. 11 Aug. 2013.

INDEX

Academic Gamers, 205
Activism, 105–6
Adler-Kassner, L., 167
Alexander, J., 6, 16, 17, 31, 66, 87, 118, 119, 120, 129, 136, 137, 139, 165, 209
Alienation, consumption and, 112
Alphabetic literacy, challenge to primacy of, 7
Alternative rhetorics, 37–38, 39, 40
  narrative and, 39
Anderson, D., 13, 106
apesmen09, 82
Assessment practices, reviews of, 52–53
Auto-ethnography, 194
Avant-garde media, 21, 87

Baca, D., 56
Bailie, B., 107, 109
Bakardjieva, M., 192
Ball, C. E., 106
Banks, A. J., 21, 56, 203
Baron, D., 39, 68
Barron, M., 193, 195
Barthes, R., 49
Bartholomae, D., 43
Bawarshi, A., 84, 85, 167, 194
Beard, J. P., 70
beautifulataxia2, 81
Beauty and the Beast, 88
La belle et la bête, 88
Benetton "Unhate" photo campaign, 110
Benjamin, W., 126
Berlin, J., 58, 105
Bey, H., 155
Bizzell, P., 38
Blakesley, D., 105, 117
Bloom, L. Z., 32, 33, 36, 37
Bolter, J. D., 55, 60
Borges, J. L., 204
Borland, J., 162
Borrowman, S., 208
Bowen, T., 44, 45
Bowlin, B., 105, 125
Brandt, D., 46, 81
Brin, D., 194
Brooke, C. G., 29, 60–61, 105, 117
Buñuel, L., 8, 9, 10
Burroughs, W. S., 204
Bush, V., 204

Cage, J., 204
Calvino, I., 204
Campbell, T., 154
Cassell, J., 207
Cell phone videos, 70
Chang, E. Y., 134–35
Cho, S.-H., 175, 176–89, 195, 208, 209
ChoLovingVampire, 185–86
"Closer" (Kim), 73, 75, 81, 103, 104
Cocteau, J., 88, 103
Colby, R., 138
Colby, R. S., 138
Collaboration, 121, 122, 128, 144–46
  authorial, 121, 122

in computer gaming, 128, 144–46
College Composition and Communication (CCC), 1, 2
"College Saga" (Leung), 171–72, 201
Communication strategies, in computer games, 140, 141–42, 148–50, 161–63
Communication studies, 36, 37, 56, 57
separation from composition, 56, 57
Composition studies, 3–7, 22–23, 24, 30–45, 167
disciplinary divide in, 36, 45
diversity in, 5
genres of, 36–37
incorporation of new media into, 3–7
lack of historicity in, 40–43
legitimacy of, 24
new media in, historical overview of, 30–45
place of formal writing in, 24
reconceptualizing, 22–23, 167
Computer games and gaming, 25, 127–70, 205–6, 207
as challenge to traditional composition, 25, 131–32
collaboration in, 128, 144–46
as compositional space, 131, 146–57
composition scholarship about, 130–40
creation of, by students, 136–37, 163–65
critical literacy development and, 128, 130–40, 146–47
dérive in, 130, 153
design of, 163–65
development of literacy strategies through, 130, 140–46
engaging student interest with, 131
gender and, 207
generational divide in knowledge of, 205–6
immersion in, 140
implications for teaching, 158–70
interaction in, 129
learning principles promoted by, 132
literacy narratives on, 158–60
metonymic dimensions of, 172
multitasking in, 141, 144
technical expertise in, 138
use in classroom, 128–29, 134–37, 158–70
utopian qualities of, 164
Computers and Composition, 31
Computers and Composition Digital Press (Utah State), 18
Computers and Writing Conference (C&W, 2009), 7–8
Conference on College Composition and Communication (CCCC), 8, 9, 10, 33, 47, 50–51, 52, 120
Statement on the Multiple Uses of Writing, 8, 9, 10, 33, 47
Consumption, spectacle and, 111–12
Council of Writing Program Administrators, 53
Couture-Nowak, J., 176
Critical literacy, 146–47
Cultural difference, mediating in *World of Warcraft,* 148–50
Cunningham, H., 207
Cushman, E., 206
Cyborg myth, 172, 189, 200

Daiker, D. A., 32, 33, 36, 37
Debord, G., 111, 112, 152
de Certeau, M., 153–54, 192
De Huergo, E., 6
DelBarrio, L., 101, 102, 103
Delivery, 29, 37
genres of, 37
Détournement, 110, 112, 113–16
queer, 113–16
Devitt, A., 194
DeVoss, D. N., 33, 34, 47, 107, 109, 110

"Digital Literacy Narrative" (beautifulataxia2), 81
"Digital Rhetoric" assignment, 97–100
Disciplinarity, 56–58
Diversity, 5
DJing, multimodal histories of, 21
Dobrin, S. I., 109
Dovey, J., 203
Downs, D., 23, 201
Dubisar, A. M., 138–39

Eason, K., 92, 93–95, 96, 97, 100
Education Arcade, 205
Emotion, and technology, 191–92
Essay, 37–38, 40, 42, 43, 78, 108
  hybrid, 40, 108
  place in composition studies, 37, 42
  reconsideration of, 43
  traditional, 38
  video, 78

Faigley, L., 36, 37, 43
Field, B., 6
Film studies, 72
Final Fantasy, 171, 172, 201
Foucault, M., 8, 9, 10, 181, 182, 188, 195
Fox, H., 38
Frasca, G., 149
Fregoso, R., 89

Games, Learning, and Society Conference, 205
Gaming. *See* Computer games and gaming
Garber, L., 115, 116
Gardiner, J., 133, 134
Gee, J. P., 25, 128, 131, 132, 133, 139, 141, 142, 144, 147, 148, 150, 151, 162, 168
Genre, 44, 85, 86, 93–96
  historical approach to, 93–96
  reshaping boundaries of, 44
  in video production, 85, 86

George, D., 6, 37
Gibbs, N., 208
Giddings, S., 203
Girard, R., 112
Girshin, T., 105, 125
Goggin, M. D., 189, 190
Goggin, P. N., 189, 190
Gould, S., 76
Granata, K., 175–76
Grant, I., 203
Gray-Rosendale, L., 37, 38, 39, 46
Groktal, 142
Gruber, S., 37, 38, 39, 46
Grusin, R., 55, 60

Handa, C., 48, 49, 108
Haraway, D. J., 172, 178, 181, 188, 193, 198, 199, 200
Harris, E., 176, 207
Hawisher, G., 6, 34, 39, 41, 42, 46, 62, 63, 67, 128, 129, 133, 139, 162, 190, 193, 204, 205
Hawk, B., 56, 57–58
Henthorne, T., 164
Hesse, D., 2, 40, 64
Hocks, M. E., 86, 106–7
Hogan, M. I., 125
Hybrid essays, 40, 108

Ideology, 18–19
Ilyasova, A., 70, 71
Images, power of, 25, 109. *See also* Photo manipulation
Interaction, multimodal, 128

Jackson, Z. A., 131, 206
Jenkins, H., 205, 207
Johansen, J., 33, 34, 35, 47
Johnson-Eilola, J., 4, 41, 42, 44, 46, 54, 55, 106, 118
Jordan, K., 204
TheJUMP (Journal for Undergraduate Media Projects), 72–73, 75, 76

Kairos, 18

Kalmbach, J., 106
Keedy, J., 49
"KEKEKE—The Ballad of Cho Seung-Hui" (ThePelly), 187
Kelly, K., 203
Kendall, L., 207
Kendrick, M. R., 86
Kesey, K., 103
Kim, K., 73–74, 75, 81, 103
King, B., 162
Kinney, K., 105, 125
Klebold, D., 176, 207
Kook-Anderson, G., 154
Koshnick, D., 167
Kress, G., 16, 17, 35, 70, 127, 131, 132, 133, 139, 142, 143, 144, 150, 151, 162, 169, 206

Lambourne, R., 187
Lanham, R. A., 174
Lashore, D., 6
Lauer, J. M., 50
Learning Games Initiative, 205
Learning outcomes, 52
Lebrescu, L., 175
Lengel, L., 207
Leonard, S. A., 46
Leung, M., 171, 172
Lieberman, M., 134, 135–36
Linder, F., 111
Lindsay F., 81–82
Lister, M., 201, 203
Literacy, 42, 77, 130–40
  computer gaming and, 130–40
  ideologies of, 77
  implications of technologies for, 42
"Literacy Narrative" (Wooten), 79–81, 82
"Literacy Narrative Project" (Lindsay F.), 81–82
Literacy practices, 34, 39
  academic versus nonstandard, 39
  cultural value and, 39
  linked to time and place, 34
  multimodal composition as, 41
Losh, E., 97–98, 100, 136, 137, 165
Lunsford, A. A., 32
Lutkewitte, C., 3, 4, 70
Lynch, D., 124
Lyotard, J.-F., 116

Mailloux, S., 56, 57
Majewski, J., 167
Manipulation of images. *See* Photo manipulation
Manovich, L., 59, 203
Mansfield, N., 181–82
Mao, L., 56
Mareck, A. F., 133, 134
Marinetti, F. T., 204
Mashups, 25, 109
Massively multiplayer online role-playing games (MMORPGs), 128, 134, 140–41, 143–44, 148–57. *See also World of Warcraft*
  communication technology used with, 140
  complex reading strategies required for, 143–44
  as compositional spaces, 150–57
  cultural difference in, mediating through communication, 148–50
  immersion in, 140
  technological literacy required for, 140–41
Mass media, consumer culture and, 173–74
McAllister, K. S., 205, 206
McCloud, S., 49
McComiskey, B., 50
McCorkle, B., 29
McIver, G., 113
McLuhan, M., 181, 204
Media, histories of, 30, 59–60
Media production, 71, 74
  anxieties about, 71
  modalities of, 74
Meeks, M., 70, 71

Micciche, L., 190, 192
Michel, A. J., 7, 20, 67, 68, 191
Miller, C., 6
Miller, S., 6, 32, 40, 56
Montfort, N., 204
Moran, C., 31
Moraski, B., 204
Morris, R. E., 48, 177, 197
Mulholland Drive, 124
Multiliteracy, 12, 13, 107
Multimedia, 4, 7, 8–13, 14–15, 19–21, 46. *See also* New media
  adoption of, 24
  composing with, 4, 13
  histories of, 20–21
  installations, 8–13
  norming of, 46
  reduction to skills, 19
  rhetorical capabilities of, 19
  student engagement with, 7, 14–15, 19–20
  student reception of, 20
  viewed through lens of literacy, 41
Multimediated selves, 171–73, 193
Multimodal composition, 1, 3–13, 16–27, 41, 44, 75, 174, 175
  agency of, 174
  colonizing pedagogies of, 13–17
  definition of, 3
  dilemmas of, 7–13
  efficacy of, 16–17
  engagement in, 22–27, 174
  as extension of traditional composition, 4
  genre boundaries and, 44
  incorporation into composition studies, 3–7
  move toward, 1
  nontextual dimensions of, 41
  reimagining, 17–22
  role in literacy education, 18
  thick network and, 175
  unpredictability in, 75
  viewed through lens of literacy, 41
Multimodal literacy, encouraging, 45
"Multimodal Literacy Project" (apesmen09), 82
Multitasking, 141, 144
Murakami, H., 15
Murray, J. M., 130, 198

Nardi, B., 150, 151, 152
Narrative, 39, 40
  alternative rhetorics and, 39
  on computer gaming, 158–60
  documentary style, 81–82
  in hybrid essays, 40
  video, 77–83
Neal, M. R., 52, 53
Newcomb, M. J., 176, 177
New Literacy, 46
New London Group, 46
New media, 13, 19, 23, 24, 30–45, 49–50, 55, 56–61, 62–69, 86–87, 97–102, 105–6, 146–47, 177, 178, 181, 192–93, 203, 205–6. *See also* Multimedia
  activism and, 105–6
  adoption of, 24
  communication possibilities of, 23
  competition with traditional print texts, 35–36
  in composition studies, historical overview of, 30–45
  critical literacy of, 146–47
  definition of, 203
  emotion and, 192–93
  expansion of composition through, 55
  generational divide in knowledge of, 205–6
  goals for, 51
  historicized view of, 46, 56–61, 68, 69, 86–87, 203
  poeticized view of, 69
  production of text with, 106–8
  prosumerism and, 105–6
  public spheres and, 62–69
  reluctance toward, 108
  rhetorical capabilities of, 19,

97–102
in service to print-based composition, 49–50
student familiarity with, 13, 204
subjectivity and, 177, 178, 181
as threat to composition, 30–34
as tool, 23

Obama, B., 101, 102

Packer, R., 204
Pain, writing about, 176–77
Palmeri, J., 5, 6, 7, 138–39
Paul, C., 136–37
Pearson, M., 204
Pegg, S., 94, 95
Penrod, D., 52
Perron, B., 205
Photo manipulation, 25, 108, 109–10, 111, 113–16, 117–26
excess and, 111, 123–26
as form of communication, 109
as generative rhetoric, 110
promiscuous, 117–23
queer *détournement* and, 113–16
Platt, J., 108, 109, 110
Porter, J. E., 29, 50
Portillo, L., 87, 88, 89, 103
Prensky, M., 205
Prosumerism, 13, 25, 106
Public spheres, 20, 58, 61, 62–69, 71
agency in, 66
complex, 64
composing in, 58, 64
importance of, 62
strategies for, 67
student participation in, 20, 67, 71

Queer, first-person iteration of, 114, 116, 119–20
Queer *détournement*, 110, 113–16
Queer theory, 115–16

Rand, E., 114, 115
Ray, S. G., 207
Reiff, M. J., 167, 194
Remediation, 55, 60
Rhetorical canons, 28–29
"Rhetoric in Practice" (RIP) assignment, 90–97
Rhodes, J., 87, 88, 117, 118
Ridolfo, J., 7, 20, 67, 68, 191
Rivers, N. A., 67
Roche, T., 208
Rodriguez, R., 82
Romano, S., 6
Romero, G., 92, 95

Sánchez, R., 173, 174
Schroeder, C., 38, 39
Sedgwick, E., 113
Selber, S., 12, 13, 21–22, 29–30, 70, 107, 146, 147, 192, 193, 196
Selfe, C. L., 1, 2–3, 4, 6, 7, 13, 33, 34, 35, 39, 41, 42, 46, 47, 49, 54, 55, 62, 63, 64, 67, 106, 107, 108–9, 118, 129, 133, 134, 139, 162, 190, 193, 204, 205
Selfe, R. L., 46, 49, 128
Shepherd, D., 6
Sheridan, D., 7, 20, 67, 68, 191
Shipka, J., 63–64
Sidler, M., 48
Sirc, G., 24, 42, 43, 46, 54, 55, 106, 117, 118, 125, 126
Situationism, 110–13, 152, 153, 154
computer gaming and, 152, 153, 154
photo manipulation and, 110
spectacle and, 111–13
Skills, reduction of technology to, 19
Smith, E. O., 48
Social media, activism and, 105–6
Socolovsky, M., 194
Song, P., 6
Sound, in composition courses, 1–2
Spectacle, 11–12, 111–13
Spooner, M., 40, 108
Stern, E., 154
Stracey, F., 112, 113

Street, B., 46
Student literacies, 46
Subjectivity, 177, 178–89, 207
Sweetland Digital Rhetoric Collaborative (University of Michigan), 18

Talking-head video, 82
Tapscott, D., 207
Techne, 116
Technical knowledge, rhetorical delivery and, 29
Technical writing, 36
Techno-comp. *See* Multimodal composition
Techno-inclusionism, 45–56, 62
Technological literacy, 140
Thebaud, J. L., 116
ThePelly, 187
Therborn, G., 18
Thick network, 175, 189–99
Third Faction, 155–57
Thurlow, C., 207
Tomic, A., 207
Toogie, 142
Transliteracies, 143–44
Trimbur, J., 37
Turing, A., 204

University of California, Irvine (UCI), Composition Program, 13–15

Vaneigem, R., 113, 117, 152
Vered, K. O., 207
Video and video production, 25, 70–105
  cell phone, 70
  composing, 86–102
  for composition courses, sample of, 78–83
  digital, 70
  genre in, 85, 86
  literacy narratives, 77–83
  place in composition classroom, 71–72
  as reenvisioning of essay, 86–102
  as replication of essay, 72–86
  rhetorical capabilities of, 73–74, 86
  technical prowess in, 84
  unpredictability in, 75, 77
  visual effects in, 88
Vielstimmig, M., 40, 41, 108
Viewmaster, 8–13, 17
Virginia Tech shootings, 25–26, 175–89, 192, 193, 207, 208–9
  disciplining of responses to, 183–84, 192, 208
  online blogs about, 177, 178–89, 193
  online game based on, 187–88
  multimedia responses to, 25–26, 175–89
  speed of response to, 181–85
  subjective normalization of, 178–89
  transmediation of, 188
  video response to, 185–87
Virilio, P., 181, 182, 195, 197
Visual field, rise of, 16, 48–49
Voice-over narration, 82

Wallace, D. L., 66
Wardle, E., 23, 201
Wardrip-Friun, N., 204
Wark, M., 152
Warner, M., 64–65, 106, 173
Warren, R., 101
"The Warrior Seung Hui Cho" (ChoLovingVampire), 185–86
Weber, R. P., 67
Weisser, C. R., 20
Welch, K., 29
Welch, N., 173
Westbrook, S., 195
White, E. M., 32, 33, 36, 37
Whithaus, C., 44, 45
Williams, J. C., Jr., 33, 34, 35, 47
Wolf, M. J. P., 205
Wolff, B., 70
Wooten, E., 79–81, 82, 83, 85, 100

Word processors, 31
World of Warcraft, 127–28, 140,141–42, 143–46, 147–57
  chat boxes in, 141
  collaborative writing and, 144–46
  communication technology used with, 140, 141–42
  complex reading strategies required for, 143–44
  as compositional space, 150–57
  cultural difference in, mediating through communication, 148–50
  discussion boards for, 145–46
  multitasking in, 141, 144
  player art exhibit, 154, 155
  racial conflict in, 147–49
  technological literacy required for, 140–41
  Third Faction and, 155–57
  transliteracies offered by, 143–44
  user guides for, 142
  website for, 142, 145–46
World Wide Web, implications for literacy, 31–32
Worsham, L., 184, 190, 191, 192, 199, 209
Writing, 36, 39
  as means of visual communication, 36
  versus nonstandard literary practices, 39
Writing Program Administrator (WPA) listserv, 2
Wysocki, A. F., 4, 41, 43, 44, 46, 54, 55, 106, 118, 122, 125

Yagelski, R. P., 46
Yancey, K. B., 2, 28, 29, 40, 108
Yee, N., 204
YouTube video project, 97–100

Zombies, course on, 92–97

AUTHORS

Jonathan Alexander is professor of English, campus writing coordinator, and director of the Center for Excellence in Writing and Communication at the University of California, Irvine. He has authored or edited eight books, including *Literacy, Sexuality, Pedagogy: Theory and Practice for Composition Studies* and the coauthored *Finding Out: An Introduction to LGBT Studies*. He is a three-time recipient of the Ellen Nold Best Article in the field of computers and composition studies. Alexander also works on several editorial boards and has been named one of the founding editorial board members for Computers and Composition Digital Press. His work focuses on the use of emerging communications technologies in the teaching of writing and in shifting conceptions of what writing, composing, and authoring mean. He also works at the intersection of the fields of writing studies and sexuality studies, where he explores what discursive theories of sexuality have to teach us about literate practice in pluralistic democracies. In 2011, Alexander received the Charles Moran Award for Distinguished Contributions to the Field of Computers and Composition.

Jacqueline Rhodes is professor of English at California State University, San Bernardino. Her scholarly work focuses on intersections of rhetoric, materiality, and technology and has been published in a variety of venues, including *College Composition and Communication*, *JAC: A Journal of Rhetoric, Culture, and Politics*, *Computers and Composition*, *Enculturation*, and *Rhetoric Review*. Her book *Radical Feminism, Writing, and Critical Agency: From Manifesto to Modem* was published in

2005, and her article "'Substantive and Feminist Girlie Action': Women Online" won the 2003 Elizabeth A. Flynn award for the most outstanding article in feminist rhetoric and composition. She serves on several editorial boards in addition to her work as interviews editor for *Composition Forum*. Grounded in her past professional life as a graphic designer and typesetter, Rhodes also focuses her energies on creative work, including multimedia installations, movies, and websites.

OTHER BOOKS IN THE CCCC STUDIES IN WRITING & RHETORIC SERIES

On Multimodality: New Media in Composition Studies
Jonathan Alexander and Jacqueline Rhodes

Toward a New Rhetoric of Difference
Stephanie L. Kerschbaum

Rhetoric of Respect: Recognizing Change at a Community Writing Center
Tiffany Rousculp

After Pedagogy: The Experience of Teaching
Paul Lynch

Redesigning Composition for Multilingual Realities
Jay Jordan

Agency in the Age of Peer Production
Quentin D. Vieregge, Kyle D. Stedman, Taylor Joy Mitchell, and Joseph M. Moxley

Remixing Composition: A History of Multimodal Writing Pedagogy
Jason Palmeri

First Semester: Graduate Students, Teaching Writing, and the Challenge of Middle Ground
Jessica Restaino

Agents of Integration: Understanding Transfer as a Rhetorical Act
Rebecca S. Nowacek

Digital Griots: African American Rhetoric in a Multimedia Age
Adam J. Banks

The Managerial Unconscious in the History of Composition Studies
Donna Strickland

Everyday Genres: Writing Assignments across the Disciplines
Mary Soliday

The Community College Writer: Exceeding Expectations
Howard Tinberg and Jean-Paul Nadeau

A Taste for Language: Literacy, Class, and English Studies
James Ray Watkins

Before Shaughnessy: Basic Writing at Yale and Harvard, 1920–1960
Kelly Ritter

Writer's Block: The Cognitive Dimension
Mike Rose

Teaching/Writing in Thirdspaces: The Studio Approach
Rhonda C. Grego and Nancy S. Thompson

Rural Literacies
Kim Donehower, Charlotte Hogg, and Eileen E. Schell

Writing with Authority: Students' Roles as Writers in Cross-National Perspective
David Foster

Whistlin' and Crowin' Women of Appalachia: Literacy Practices since College
Katherine Kelleher Sohn

Sexuality and the Politics of Ethos in the Writing Classroom
Zan Meyer Gonçalves

African American Literacies Unleashed: Vernacular English and the Composition Classroom
Arnetha F. Ball and Ted Lardner

Revisionary Rhetoric, Feminist Pedagogy, and Multigenre Texts
Julie Jung

Archives of Instruction: Nineteenth-Century Rhetorics, Readers, and Composition Books in the United States

Jean Ferguson Carr, Stephen L. Carr, and Lucille M. Schultz

Response to Reform: Composition and the Professionalization of Teaching
Margaret J. Marshall

Multiliteracies for a Digital Age
Stuart A. Selber

Personally Speaking: Experience as Evidence in Academic Discourse
Candace Spigelman

Self-Development and College Writing
Nick Tingle

Minor Re/Visions: Asian American Literacy Narratives as a Rhetoric of Citizenship
Morris Young

A Communion of Friendship: Literacy, Spiritual Practice, and Women in Recovery
Beth Daniell

Embodied Literacies: Imageword and a Poetics of Teaching
Kristie S. Fleckenstein

Language Diversity in the Classroom: From Intention to Practice
edited by Geneva Smitherman and Victor Villanueva

Rehearsing New Roles: How College Students Develop as Writers
Lee Ann Carroll

Across Property Lines: Textual Ownership in Writing Groups
Candace Spigelman

Mutuality in the Rhetoric and Composition Classroom
David L. Wallace and Helen Rothschild Ewald

The Young Composers: Composition's Beginnings in Nineteenth-Century Schools
Lucille M. Schultz

Technology and Literacy in the Twenty-First Century: The Importance of Paying Attention
Cynthia L. Selfe

Women Writing the Academy: Audience, Authority, and Transformation
Gesa E. Kirsch

Gender Influences: Reading Student Texts
Donnalee Rubin

Something Old, Something New: College Writing Teachers and Classroom Change
Wendy Bishop

Dialogue, Dialectic, and Conversation: A Social Perspective on the Function of Writing
Gregory Clark

Audience Expectations and Teacher Demands
Robert Brooke and John Hendricks

Toward a Grammar of Passages
Richard M. Coe

Rhetoric and Reality: Writing Instruction in American Colleges, 1900–1985
James A. Berlin

Writing Groups: History, Theory, and Implications
Anne Ruggles Gere

Teaching Writing as a Second Language
Alice S. Horning

Invention as a Social Act
Karen Burke LeFevre

The Variables of Composition: Process and Product in a Business Setting
Glenn J. Broadhead and Richard C. Freed

Writing Instruction in Nineteenth-Century American Colleges
James A. Berlin

Computers & Composing: How the New Technologies Are Changing Writing
Jeanne W. Halpern and Sarah Liggett

A New Perspective on Cohesion in Expository Paragraphs
Robin Bell Markels

Evaluating College Writing Programs
Stephen P. Witte and Lester Faigley

This book was typeset in Garamond and Frutiger by Barbara Frazier.
Typefaces used on the cover include Adobe Garamond and Formata.
The book was printed on 55-lb. Natural Offset paper
by Versa Press, Inc.

www.ingramcontent.com/pod-product-compliance
Lightning Source LLC
Chambersburg PA
CBHW071419050326
40689CB00010B/1900

* 9 7 8 0 8 1 4 1 3 4 1 2 2 *